General Relativity and Cosmology:

A First Course

Tai L Chow
California State University, Stanislaus

Wuerz Publishing Ltd.
Winnipeg, Canada

phone (204) 453 7429
fax (204) 453 6598

Wuerz Publishing Ltd.
Winnipeg, Canada

General Relativity And Cosmology: A First Course
Tai L Chow

ISBN 0-920063-54-3

Printed in Canada

To the memory of my parents,

周建侯 祁健

Chow Jiann Hour Chi Jiann

Preface

Many people are first excited in physics by popular books or films about cosmology. But the subject has been rarely taught in universities. Perhaps this is partly because a suitable textbook has not been available. There are many advanced books on the theory of general relativity and cosmology for the specialist, and many elementary expositions for the lay reader. What a pity, then, that there is a gap at the undergraduate level. This book is an attempt to fill the gap. We try to make available to the student a working acquaintance with the concepts, fundamental ideas and modes of calculation used in general relativity, so the student can continue more deeply into any specialized aspects of the subject which interest him/her.

This book evolved from a set of lecture notes for a course which I have taught over the past ten years. I assume that the student has been exposed to a calculus-based course in general physics from a text such as Resnick & Halliday, and a course in calculus (including the handling of differentiations of field equations). Some exposure to tensor analysis would be helpful, but not necessary; this subject is covered in the text.

The student will find that in the derivations of equations, a generous amount of detail has been given. However, to ensure that the student does not lose sight of the development underway, some of the more lengthy and tedious algebraic manipulations have been omitted when necessary.

Tai L. Chow
Department of Physics
California State University, Stanislaus
Turlock, California 95380

Introduction

In the early 1900s, three events took place in physics which dramatically changed the course of modern physics. In 1905 Einstein formulated the Special Theory of Relativity, then in 1915 he developed the General Theory of Relativity, and around 1925 quantum mechanics took its present form. Since then physics has progressed rapidly. Beginning in 1930, quantum mechanics and special relativity were united into relativistic quantum field theory. This merger was very rewarding in that it gives, at the least, partial explanation of the laws and interactions governing elementary particle physics.

Among the four types of forces (strong, electromagnetic, weak, and gravitational) known today, gravity is a very strange force. Weak though it is, gravity has two properties which make it dominate the other three forces over cosmic distances: it is a long-range force, and it cannot be screened. Therefore any cosmology must be founded on a logically secure theory of gravitation.

The first three forces could be explained through particle interactions taking place in the flat space-time of special relativity. However, gravity defies such an explanation. In order to describe the mysterious gravity, Einstein in 1915 was compelled to generalize the ideas of his special relativity and he eventually connected gravity with the geometry of space-time. So Einstein's general theory of relativity is, as well, a relativistic theory of gravitation.

For a long time, Einstein's theory of general relativity occupied an isolated position within the domain of general physics. This was attributable in part to the mathematical framework of the theory, which is based on Riemannian geometry (tensor calculus) which is not needed in other physical applications. The extreme difficulty in devising suitable experiments which might verify the theory, and the growth of more fertile fields of investigation such as atomic and nuclear physics, elementary particles also contributed to the isolation of the theory. However, Einstein's theory of general relativity is now meeting renewed interest. This is due partly to the development of new technological capabilities which opened up previously inaccessible avenues for the experimental verification of general relativity, and partly to the conjecture of some theoretical physicists that the fundamental difficulties confronting quantum field theory may find their resolution in a suitable combination of the two disciplines. The discovery of compact celestial objects - neutron stars for instance, or black holes - provides the final turning point. It demands the application of Einstein's theory of general relativity for the study of these objects. Today, physics and astronomy have joined forces to form the discipline called relativistic astrophysics. Einstein's general relativity also has a great deal to say about cosmology, since according to Einstein's theory, the overall space-time structure is intimately related to the gravitational field (via the metric tensor).

There are many advanced books on theory of general relativity and cosmology, but few at the undergraduate level. In this book, I try to make available to the student a working familiarity with the concepts, fundamental ideas and calculational techniques used in general relativity, to the extent that he/she can appreciate the mainstream of modern literature on general relativity, ie, relativistic theory of gravitation and cosmology.

CONTENTS

Chapter 3 Basic Ideas of General Relativity 61

Chapter 4 Curvilinear Coordinates and General Tensors 81

Chapter 5 Einstein's Law of Gravitation 113

Chapter 1

Classical Mechanics

Classical mechanics studies the motion of physical bodies at the macroscopic level. Its foundations were first laid in 17th century by Newton and Galileo. The essential physics of the mechanics of Newton & Galileo, known as Newtonian mechanics, is contained in the three laws of motion. Classical mechanics has since been reformulated in a few different forms: the Lagrange, the Hamilton, and the Hamilton-Jacobi formalisms. They are alternatives, but equivalent to the Newtonian mechanics. We will review Newtonian mechanics first and then Lagrangian dynamics.

1-1 Newtonian mechanics

A. The Three Laws of motion

The foundation of the Newtonian mechanics is found in the three laws of motion which were postulated by Newton as a result of the combination of experimental evidence and a great deal of intuition. Recall these three laws:

(1) The First Law (also known as Galileo's law) states: A free particle continues in its state of rest or of uniform motion until it is acted upon by an external force.

Newton made the First Law more precise by introducing the concepts of "quantity of motion" and "amount of matter" which we now call momentum, and mass of the particle, respectively. The momentum $\vec{\mathbf{p}}$ of a particle is related to its velocity $\vec{\mathbf{v}}$ and mass m by the relation

$$\vec{p} = m\vec{v}. \qquad (1)$$

The First Law may now be expressed mathematically as

$$\vec{p} = m\vec{v} = \text{constant} \tag{2}$$

in the absence of external force acting on the particle. Thus, the First Law is a Law of Conservation of Momentum.

(2) *The Second Law gives a specific way of determining how the state of motion of a particle is changed when it is acted on by a force:*

The time rate of change of momentum of the particle is equal to the external applied force

$$\vec{F} = d\vec{p}/dt \; . \tag{3}$$

Since $\vec{\mathbf{p}} = m\vec{\mathbf{v}}$, and if the mass m is constant, then the Second Law can also be written as

$$\vec{\mathbf{F}} = m\, d\vec{\mathbf{v}}\,/\,dt = m\vec{\mathbf{a}} \tag{4}$$

which says that the force acting on a particle (of constant mass) is equal to the product of its acceleration and its mass, a familiar result from basic physics.

Mass, which appears in Newton's laws of motion, is called the inertial mass of the particle, as it is indicative of the resistance offered by the particle to a change of its velocity (i.e., it is the measure of the particle's inertia).

(3) *The Third Law states that if a particle A exerts a force* $\vec{\mathbf{F}}$ *on a particle B, then B exerts a force* $-\vec{\mathbf{F}}$ *on particle A.*

The Third Law is often simply stated as: "The forces of action and reaction are equal and opposite."

Physical laws are usually stated relative to some reference frame, which allows physical quantities like velocity, acceleration, etc., to be defined. Newton's laws apply in **inertial frames**. An inertial frame of reference is a reference frame in which the free motion of any body takes places with a constant velocity. It is also a basic assumption of classical mechanics that we can introduce or associate a Cartesian coordinate system S with such a frame, and the geometry in such a frame is **Euclidean**. Also, a time coordinate t must be introduced, or defined separately, in order that the Cartesian coordinate system can be used to catalog events.

B. The Galilean transformation

Physical experiments can be carried out in different frames of reference. We must know how to convert the results of measurement obtained in one reference frame into those which are obtained in another. Now consider two frames S and S′ which coincide at time $t = t' = 0$, as shown in Fig. 1: S′ moves with a constant velocity $\vec{\mathbf{V}}$ relative to S. The corresponding axes of S and S′ remain parallel throughout the motion. It is assumed that the same units of distance and time are adopted in both frames. An inspection of Figure 1 gives

$$\vec{\mathbf{r}}' = \vec{\mathbf{r}} - \vec{\mathbf{V}}t, \qquad \text{and} \qquad t' = t \tag{5}$$

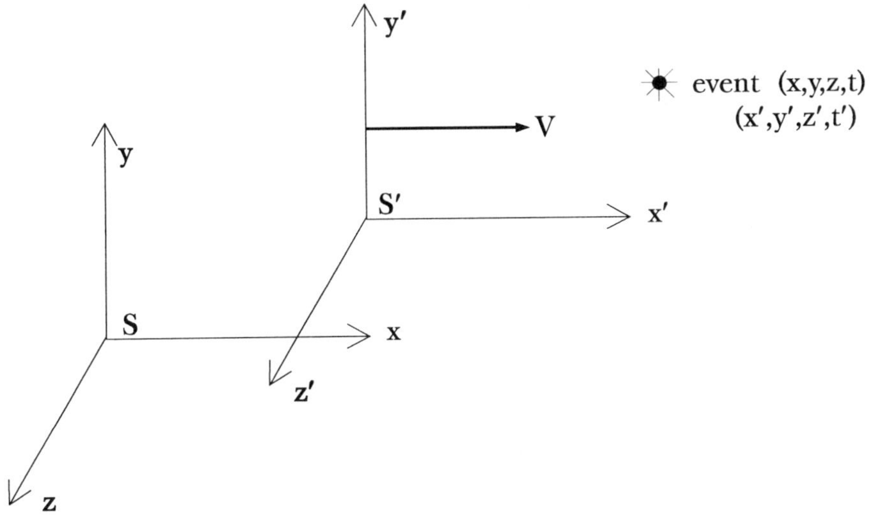

Figure 1: Coordinate frames S and S′ moving with relative speed v.

These transformations are called the **Galilean transformations** (GT). The first relation of Eq.(5) is commonly called a **boost**, and the second relation, $t' = t$, is usually taken for granted. It asserts that time is not affected by relative motion - i.e., it expresses the universality, or absoluteness of time: time flows absolutely, uniformly, and evenly, without relationship to the actual events that occur as time passes.

We shall see later that the Galilean transformations are only valid for velocities which are small compared with that of light.

Differentiating Eq.(5) once, we obtain

$$\dot{\vec{\mathbf{r}}}' = \dot{\vec{\mathbf{r}}} - \vec{\mathbf{V}} \tag{6}$$

which says that if the velocity $\dot{\vec{\mathbf{r}}}$ of a particle in the frame S is a constant, then its velocity $\dot{\vec{\mathbf{r}}}'$ in the frame S′ is also a constant. Differentiating the velocity relation once with respect to time, we obtain

$$\ddot{\vec{\mathbf{r}}}' = \ddot{\vec{\mathbf{r}}} \tag{7}$$

which indicates that the acceleration of the particle is also a Galilean invariant.

If the frame S in Figure 1 is inertial, so is S′, since the linear equations of motion of *free* particles in frame S are transformed by Eq.(5) into similar linear equations in S′. Any frame which moves *uniformly* (i.e., with constant velocity and without rotation) relative to any inertial frame is also itself an inertial frame. And there is an infinity of inertial frames, all connected by Galilean transformations.

C. Newtonian relativity & Newton's absolute space

The inertial mass m is also a Galilean invariant, so

$$\mathbf{m}\ddot{\vec{\mathbf{r}}}' = \mathbf{m}\ddot{\vec{\mathbf{r}}}$$

that is, the $\mathbf{m}\vec{\mathbf{a}}$ term of Newton's Second Law is a Galilean invariant. The applied force on the particle is also invariant under Galilean transformations provided that it is velocity-independent. That is, if the applied force is velocity-independent, the form of Newton's Second Law is preserved under Galilean transformations. Thus, not only Newton's First Law, but also his Second and Third Laws, are valid in all inertial frames. This property of classical mechanics is often referred to as **Newtonian (or Galilean) relativity**.

Because of this relativity, the uniform motion of one inertial frame relative to another cannot be detected by internal mechanical experiments subject to Newton's theory. According to the First Law, a particle does not resist uniform motion, of whatever speed, but it does resist any change in its velocity, i.e., acceleration. This is precisely expressed by Newton's Second Law and the mass m is a measure of the particle's inertia. Here we should ask: "acceleration with respect to what?" One may give a simple answer: with respect to any one of the inertial frames. This answer is quite unsatisfactory. Why does nature single out such "preferred" frames as standards of acceleration? Inertial frames are unaccelerated and non-rotating, but relative to what? Newton found no answer, and postulated instead the existence of an **absolute space**, which "exists in itself, as if it were a substance, with basic properties and quantities that are not dependent on its relationship to anything else whatsoever (e.g., the matter that is in this space)". Newton's concept of an absolute space has never lacked critics. From Huyghens, Leibniz, Bishop Berkeley, to Ernst Mach (in the 19th century) and Einstein (in the 20th century), serious objections have been brought against absolute space:

(1) It is purely *ad hoc* and explains nothing. It is similar to Kepler's early suggestion that angles were responsible for pushing the planets along their paths.

(2) How are we to identify which inertial frame is at rest relative to absolute space?

(3) Newton's absolute space is a physical entity, thus it acts on matter (it is the "seat of inertia" resisting acceleration in the absence of forces). However, matter does not act on it (it is "without relation to anything external"). As Einstein said: "It conflicts with one's scientific understanding to conceive of a thing (absolute space) which acts, but cannot be acted upon."

This is perhaps the most powerful objection. It questions not only absolute space (which Newton's theory can do without) but also the set of all inertial frames.

Strictly speaking, it is impossible to determine whether or not a given frame is an inertial frame, because there are at least some gravitational forces acting unless the body was removed infinitely far away from all other matter. Therefore, in practice we merely specify an approximate inertial frame in accordance with the needs of the problem under investigation. For example, our earth is a good approximation to an inertial frame for many terrestrial applications; and the sun, at rest relative to the average motion of nearby stars, provides a better approximation to an inertial frame which enables us to use Newtonian mechanics to describe motion within the solar system with greater accuracy. The center of our galaxy, which appears at rest relative to the average motion of other galaxies, is an even better approximation to an inertial frame; in this frame the dynamics of our galaxy can be understood within Newtonian mechanics. All this experimental evidence indicates that Newton's absolute space seems not to be absolute at all, but somehow tied to the distribution of matter on a large scale.

D. Newtonian gravity

Historically, Newton began with his law of motion and Kepler's elliptical orbits to arrive at his empirical law of gravity between two masses m_1 and m_2 separated by a distance r (Figure 2)

$$\vec{F} = -\frac{Gm_1m_2}{r^2}\hat{r} \quad . \tag{8}$$

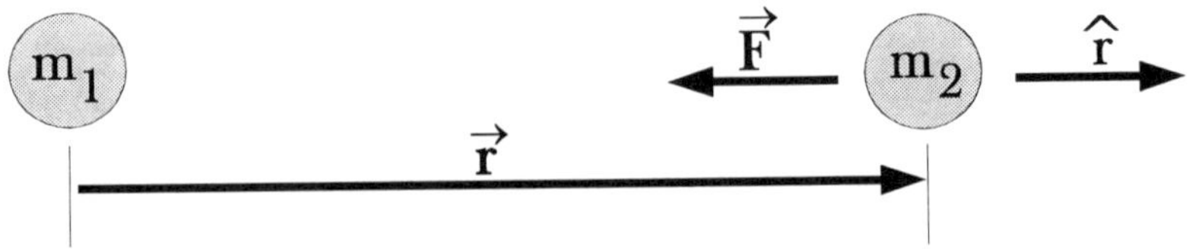

Figure 2: Gravitational force on man m_2 due to man m_1.

According to this law, gravitational forces act at a distance and are able to produce their effects through millions of miles of empty space.

The Galilean transformation preserves the distance between two points in space, thus Newton's law of gravity is invariant under the Galilean transformation, and the gravitational influence propagates with infinite speed.

Gravity is a very strange force in Newtonian mechanics. Consider Newton's Second Law

$$\vec{F} = m\vec{a} \quad . \tag{9}$$

$\vec{F}$ arises from the action of other matter on m, and may be electric, frictional, gravitational etc. Now let $\vec{F}$ be a gravitational force, arising from N masses m_1, m_2, ..., m_N at fixed positions $\vec{r}_1, \vec{r}_2, ... \vec{r}_N$ (Figure 3). If $\vec{r}(t)$ is the changing position of m, then Newton's law of gravity gives

$$\vec{F} = -\sum_{i=1}^{N} \frac{G m m_i (\vec{r} - \vec{r}_i)}{|\vec{r} - \vec{r}_i|^3} \tag{10}$$

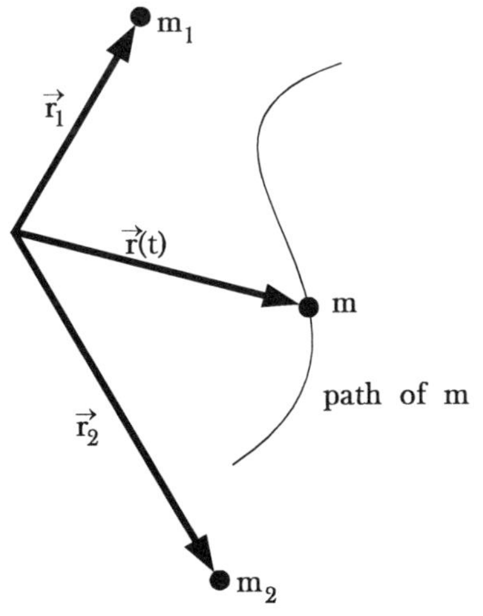

Figure 3: Path of a particle.

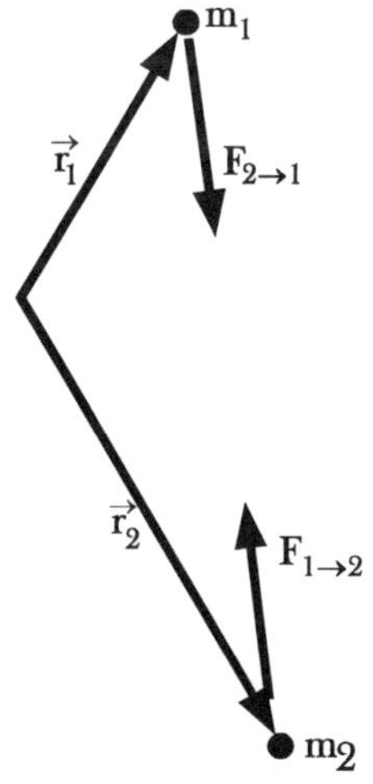

Figure 4: Action and reaction.

from which we see that the mass m which governs a body's resistance to external applied force — the inertial mass — is also the mass which determines the magnitude of the gravitational force acting on it — i.e. the passive gravitational mass. Gravitational mass m_G occurs in Newton's law of gravity in two ways: passively (m_G^p) in determining the force on the body from others, and actively (m_G^a) in determining the force with which the body attracts others. Active and passive gravitational masses are proportional, and without loss of generality we can define them as equal. This follows from the Third Law of motion. Consider two particles 1 and 2, at $\vec{\mathbf{r}}_1$ and $\vec{\mathbf{r}}_2$ (Figure 4). The force $\vec{\mathbf{F}}_{2\to 1}$ that 2 exerts on 1 is

$$\vec{\mathbf{F}}_{2\to 1} = - G m_{G1}^p \, m_{G2}^a (\vec{\mathbf{r}}_1 - \vec{\mathbf{r}}_2) \, / \, |\vec{\mathbf{r}}_1 - \vec{\mathbf{r}}_2|^3$$

while

$$\vec{\mathbf{F}}_{1\to 2} = - G m_{G2}^p \, m_{G1}^a (\vec{\mathbf{r}}_2 - \vec{\mathbf{r}}_1) \, / \, |\vec{\mathbf{r}}_1 - \vec{\mathbf{r}}_2|^3$$

is the force that 1 exerts on 2. The Third Law of motion then gives

$$m_{G1}^p \, m_{G2}^a = m_{G2}^p m_{G1}^a \qquad \text{or} \qquad m_{G1}^p / m_{G2}^p = m_{G1}^a / m_{G2}^a ,$$

that is, active and passive gravitational masses are proportional.

This dual role played by mass has an astonishing consequence. Inserting Eq.(10) into Eq.(9), we see that m cancels and we get

$$\vec{a} = -G \sum_{i=1}^{N} \frac{m_i(\vec{r} - \vec{r}_i)}{|\vec{r} - \vec{r}_i|^3}$$

i.e., the acceleration of a particle in any gravitational field is independent of its mass. Thus, if only gravitational forces act, all bodies similarly projected pursue identical trajectories. This was known to Galileo and Newton. Both Galileo and Newton conducted experiments to test the equivalence of inertial and gravitational masses, and their experiments suggest that

$$m_I = m_G \, . \tag{11}$$

It has been established by Roll, Krotkov and Dicke in 1964, and by Panov and Braginski in 1971 that they are equal within one part in 10^{12}.

The assertion of the equivalence of the inertial mass and gravitational mass is known as the **principle of equivalence**. Pursuit of the consequences of the principle of equivalence led Einstein to formulate his theory of general relativity between 1911 and 1916.

E. Gravitational field

We now attempt to reformulate Newton's theory of gravitation so that action at a distance is eliminated. This can be done very easily through the field concept. Here is the basic idea: consider any body - the sun, for example, in space. The presence of this body alters the properties of the space in its vicinity, setting up a gravitational field which stands ready to interact with any other masses brought into it. We now try to make these ideas quantitative

The gravitational field intensity **g** at any point in space may be defined as the gravitational force per unit mass which would act on a particle located at that point,

$$\vec{\mathbf{g}} = \vec{\mathbf{F}}/m \,. \tag{12}$$

Thus at a distance r from a body of mass M, a particle of mass m would experience a force

$$\vec{F} = -\frac{GMm}{r^2}\hat{r}$$

where $\hat{r}$ is a unit vector away from M. Therefore the intensity of the gravitational field at a distance r from M is

$$\vec{g} = \frac{\vec{F}}{m} = -\frac{GM}{r^2}\hat{r} \,. \tag{13}$$

If more than one body is present, the gravitational field is the vector sum of the individual fields produced by each body. For a body which consists of a continuous distribution of matter, Eq. (13) becomes

$$\vec{g} = -G\int_{vol} \frac{\rho(\vec{r}')\hat{r}}{r^2} dV' \quad , \tag{14}$$

Upon substituting the identity $\nabla(1/r) = -(1/r^2)\,\hat{r}$ into Eq.(14), we find that

$$\vec{g} = G\int_{vol} \nabla\left(\frac{1}{r}\right)\rho(\vec{r}')dV' \ ,$$

Since ∇ does not operate on the variable $\vec{r}'$, it can be factored out of the integral and we have

$$\vec{g} = \nabla\int_{vol} \frac{G\rho(\vec{r}')}{r} dV' \quad , \tag{15}$$

Eq.(15) indicates that $\vec{g}$ can be represented as the gradient of a scalar function ϕ:

$$\vec{g} = -\nabla\Phi \tag{16}$$

with

$$\phi = -\int_{vol} \frac{G\rho(\vec{r}')}{r} dV' = -\frac{GM}{r} \tag{17}$$

Thus, the static gravitational field is a conservative vector field. The scalar function ϕ is called the **gravitational potential** and has dimension of energy per unit mass. (It is the gravitational potential energy per unit mass.)

F. Gravitational field equations

It is of interest to find differential equations for the gravitational potential ϕ. To this end, we consider a closed surface S enclosing a mass M (Figure 5). The quantity $\vec{g}\cdot\hat{n}dS$ is defined as the gravitational flux passing through the surface element dS, where $\hat{n}$ is the outward unit vector normal to dS. The total gravitational flux through S is then given by

$$\int_S \vec{g}\cdot\hat{n}\,dS = -\ GM\int_S \frac{\hat{r}\cdot\hat{n}}{r^2}dS \tag{18}$$

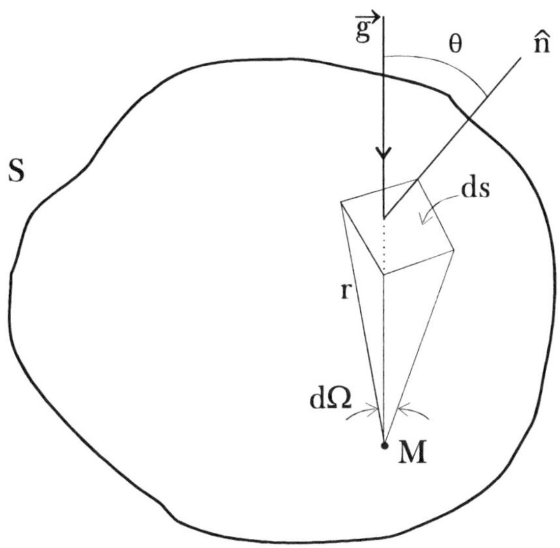

Figure 5: The element of solid angle dΩ is defined as $(\hat{r}\cdot\hat{n}/r^2)\,ds$.

By definition $(\hat{r}\cdot\hat{n}/r^2)dS = \cos\theta\ dS/r^2$ is the element of solid angle dΩ subtended at M by the element of surface dS. And so Eq.(18) becomes

$$\int_S \vec{g}\cdot\hat{n}\,dS = -GM\int d\Omega = -4\pi GM \ . \tag{19}$$

If the surface S encloses a number of masses M_i, then

$$\int_S \vec{g}\cdot\hat{n}\,dS = -4\pi G\sum_i M_i \quad . \tag{20}$$

For a continuous distribution of mass within S, the sum becomes an integral

$$\int_S \vec{g}\cdot\hat{n}\,dS = -\int_V 4\pi G\rho(\vec{r})dV \quad . \tag{20a}$$

The left side of this equation can be rewritten, with the application of Gauss' divergence theorem

$$\int_S \vec{g}\cdot\hat{n}\,dS = \int_V \nabla\vec{g}\,dV \quad .$$

Substituting this into Eq.(20a), we have $\int_V (\nabla\cdot\vec{g} + 4\pi G\rho)dV = 0.$

This equation holds true for any volume and it can be true only if the integrand vanishes:

$$\nabla\cdot\vec{g} + 4\pi G\rho = 0 \tag{21}$$

and recall that

$$\vec{g} = -\nabla\phi \quad . \tag{22}$$

So we finally arrive at the result:

$$\nabla^2\phi(\vec{r}) = 4\pi G\rho(\vec{r}) \tag{23}$$

which is called Poisson's equation. If $\rho = 0$, Poisson's equation reduces to Laplace's equation

$$\nabla^2\phi(\vec{r}) = 0 \quad . \tag{24}$$

Equations (23) and (24) constitute the field equations for the Newtonian theory of gravitation.

1-2 Lagrangian mechanics

As we mentioned in the beginning of this chapter, classical mechanics can be reformulated in a few different forms - such as the Lagrange, the Hamiltonian, and the Hamilton-Jacobi formalisms. Each is based on the ideas of work or energy, and each is expressed in terms of *generalized coordinates*. Any convenient set of parameters (quantities) which can be used to specify the state of the system can be taken as the generalized coordinates. Thus, the new generalized coordinates may be any quantities that can be observed to change with the motion of the system, and they need not be geometrical quantities. They may be electric charges, for example, in certain circumstances. We shall write the generalized coordinates as q_i, i = 1,2,3,...n, where n is the number of degrees of freedom of the system.

In terms of these generalized coordinates q_i, we can write the basic equations of motion in some form that is equally suitable for all coordinate systems, and they can be applied to a wide range of physical phenomena, particularly those involving fields with which Newton's equations of motion are not usually associated.

We present one of the alternative forms here: the Lagrangian dynamics, developed by Joseph Louis Lagrange (1736-1813). Lagrange selected D'Alembert's principle as the starting point and obtained the equations of motion "Lagrange's equations" from it. Instead of D'Alembert's principle, we will start with a variational principle, known as Hamilton's principle.

Hamilton's principle may be stated as follows:

Of all possible paths along which a dynamical system may move from one point to another within a specific time interval and consistent with any constraints, the actual path followed is that for which the time integral of the difference between the kinetic and potential energies has stationary value.

In terms of the calculus of variation, Hamilton's principle becomes

$$\delta\int_{t_1}^{t_2}(T - V)dt = \delta\int_{t_1}^{t_2}L\ dt = 0$$

where L is defined to be the difference between the kinetic and potential energies and is called the Lagrangian of the system. Energy is a scalar quantity and so the Lagrangian is a scalar function. Hence the Lagrangian must be invariant with respect to coordinate transformations. We are therefore assured that no matter what generalized coordinates are chosen for the description of a system, the Lagrangian will have the same value for a given condition of the system. Although Lagrangian will be expressed by means of different functions depending on the generalized coordinates used, the value of the Lagrangian is unique for a given condition. Therefore, we can write

$$L = L(q_i, \dot{q}_i, t) = T(q_i, \dot{q}_i, t) - V(q_i, t) \tag{25}$$

where $\dot{q}_i$ is the generalized velocity corresponding to q_i. Hamilton's principle becomes

$$\delta I = \delta\int_{t_1}^{t_2}L(q_i, \dot{q}_i, t)dt = 0 \tag{26}$$

where $q_i(t)$, and hence $\dot{q}_i(t)$, is to be varied subject to the end conditions: $\delta q_i(t_1) = \delta q_i(t_2) = 0$. The symbol "$\delta$" refers to the variation in a quantity between two paths, and "d" as usual refers to a variation along a given path. If we label each possible path by a parameter, δ then stands as a shorthand for the parametric procedure outlined below.

We label each possible path by a parameter α in the following way:

$$q_i(t,\alpha) = q_i(t,0) + \alpha\eta_i(t) \tag{27}$$

where $q_i(t,0)$ is the actual dynamical path followed by the system (as yet unknown), and $\eta_i(t)$ is a completely arbitrary function of t which has a continuous first derivative and subject to $\eta_i(t_1) = \eta_i(t_2) = 0$. In terms of the variation symbol, we can write

$$\delta q_i = \frac{\partial q_i}{\partial \alpha} d\alpha = \eta_i d\alpha \tag{28}$$

which corresponds to a virtual displacement in q_i from the actual dynamical path to a neighboring varied path. The situation is depicted schematically in Figure 6:

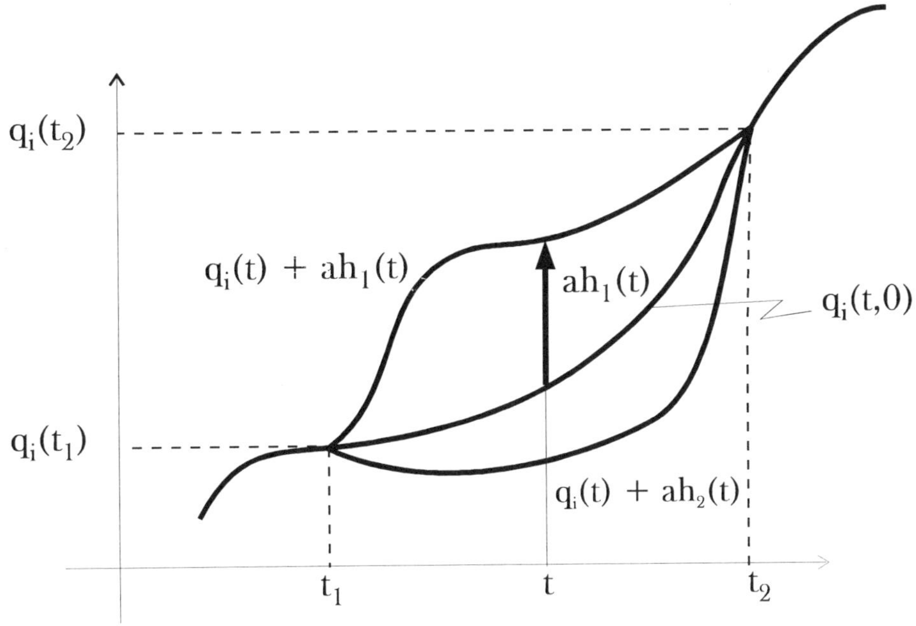

Figure 6: Illustrating the variation of q_i.

The action integral I is now a function of α only, for any given $\eta_i(t)$:

$$I(\alpha) = \int_{t_1}^{t_2} L(q_i(t,\alpha),\dot{q}_i(t,\alpha),t)dt \quad , \tag{29}$$

Hence the stationary values of $I(\alpha)$ occur when $\partial I/\partial \alpha = 0$. But by our choice of $q_i(t,0)$, we know that this occurs when $\alpha = 0$, so that the necessary condition (that the action integral has a stationary value) is $\partial I/\partial \alpha = 0$ when $\alpha = 0$. In terms of the variation symbol δ, this necessary condition can be written as

$$\delta I = \left(\frac{\partial I}{\partial \alpha}\right)_{\alpha=0} d\alpha = 0 \tag{30}$$

for arbitrary $\eta_i(t)$ and nonzero α, and the subscript $\alpha = 0$ means that we evaluate the derivative $\partial I/\partial\alpha$ at $\alpha = 0$.

We now expand the integrand L in Eq.(29) in a Taylor's series

$$I(\alpha) = \int_{t_1}^{t_2}[L(q_i(t,0),\dot{q}_i(t,0);t) + \alpha\eta_i\frac{\partial L}{\partial q_i} + \alpha\dot{\eta}_i\frac{\partial L}{\partial \dot{q}_i} + 0(\alpha^2)]dt\ . \tag{31}$$

Since the integration limits t_1 and t_2 are not dependent on α, we can differentiate under the integral sign with respect to α and obtain

$$\frac{\partial I}{\partial \alpha} = \int_{t_1}^{t_2}\left[\frac{\partial L}{\partial q_i}\eta_i + \frac{\partial L}{\partial \dot{q}_i}\dot{\eta}_i + 0(\alpha^2)\right]dt\ . \tag{32}$$

Dropping terms in α^2, α^3, etc., and integrating by parts the second term, yields

$$\int_{t_1}^{t_2}\frac{\partial L}{\partial \dot{q}_i}\dot{\eta}_i dt = \left[\frac{\partial L}{\partial \dot{q}_i}\eta_i(t)\right]_{t_1}^{t_2} - \int_{t_1}^{t_2}\eta(t)\frac{d}{dt}\left(\frac{\partial L}{\partial \dot{q}_i}\right)dt\ . \tag{33}$$

The first term on the right is zero because $\eta_i(t_1) = \eta_i(t_2) = 0$. Substituting Eq.(33) into Eq.(32) we obtain

$$\left(\frac{\partial I}{\partial \alpha}\right)_{\alpha=0} = \int_{t_1}^{t_2}\left[\frac{\partial L}{\partial q_i} - \frac{d}{dt}\left(\frac{\partial L}{\partial \dot{q}_i}\right)\right]_{\alpha=0}\eta_i(t)dt\ . \tag{34}$$

To obtain the stationary condition in terms of the variation symbol δ, we multiply Eq.(34) through by $d\alpha$ resulting in

$$\left(\frac{\partial I}{\partial \alpha}\right)_{\alpha=0} d\alpha = \int_{t_1}^{t_2}\left[\frac{\partial L}{\partial q_i} - \frac{d}{dt}\left(\frac{\partial L}{\partial \dot{q}_i}\right)\right]d\alpha\eta_i(t)dt \tag{35}$$

$$\delta I = \int_{t_1}^{t_2}\left[\frac{\partial L}{\partial q_i} - \frac{d}{dt}\left(\frac{\partial L}{\partial \dot{q}_i}\right)\right]\delta q_i(t)dt \quad . \tag{35}$$

Since $\delta q_i(t)$ is arbitrary except when $\delta q_i(t_1) = \delta q_i(t_2) = 0$, it follows that a necessary condition for $\delta I = 0$ is that the square bracket vanishes, yielding Lagrange's equations

$$\frac{d}{dt}\left(\frac{\partial L}{\partial \dot{q}_i}\right) - \frac{\partial L}{\partial q_i} = 0, \qquad i = 1,2,3,....n \quad . \tag{36}$$

This is also a sufficient condition for a stationary value of the action integral I. The sufficient condition is apparent from the fact that Eq. (36) implies that the integral (34) vanishes, resulting in the variation δI being zero.

Problems

(1-1) Under the influence of a force field a particle of mass m moves in the xy-plane and its position vector is given by

$$\vec{r} = a\cos\omega t\,\hat{i} + b\sin\omega t\,\hat{j}$$

where a, b and ω are positive constants and $a > b$.

(a) Show that the particle moves in an ellipse.

(b) Show that the force acting on the particle is always directed toward the origin.

(c) Show that

$$\vec{r}\times\vec{p} = mab\omega\,\hat{k}$$
$$\vec{r}\cdot\vec{p} = \frac{1}{2}m(b^2 - a^2)\sin 2\omega t$$

where $\vec{p}$ is the momentum of the particle.

(1-2) Two astronauts of masses M_A and M_B, initially at rest in free space, pull on either end of a rope. The maximum force with which astronaut A can pull, F_A, is larger than the maximum force with which astronaut B can pull, F_B. Find their motion if each pulls on the rope as hard as possible.

(1-3) Show that the electromagnetic wave equation

$$\frac{\partial^2\phi}{\partial x^2} + \frac{\partial^2\phi}{\partial y^2} + \frac{\partial^2\phi}{\partial z^2} - \frac{1}{c^2}\frac{\partial^2\phi}{\partial t^2} = 0$$

is not invariant under the Galilean transformations.

(1-4) (a) Find the force of attraction of a thin spherical shell of radius a on a particle P of mass m at a distance $r > a$ from its center.

(b) Prove that the force of attraction is the same as if all the mass of the spherical shell were concentrated at its center.

(1-5) Derive the result of Problem 1-4a by first finding the potential due to the mass distribution.

(1-6) A simple pendulum of mass m and string length b oscillates in a plane about its equilibrium position. If θ is the angular displacement of the pendulum from equilibrium ($\theta = 0$),

(a) write the Lagrangian of the system in terms of θ and b,

(b) use the Lagrange's equation to show that

$$\ddot{\theta} + \frac{g}{b}\sin\theta = 0 \ .$$

Chapter 2

Special Relativity

2-1 Historical origins of special relativity

The Galilean transformation asserts that any one inertial frame is as good as another for describing the laws of classical mechanics. But physicists of the 19th century were unable to grant the same freedom to electromagnetic theory, which is not invariant under the Galilean transformation. The classical electromagnetic theory is summarized in Maxwell's equations, and one of the consequences of Maxwell's equations is that speed of light is independent of the motion of the light source. However, it can be demonstrated that, under the Galilean transformation, in a frame of reference moving uniformly with respect to the light source, the light wave is no longer spherical and the speed of light is also different. For electromagnetic phenomena, therefore, inertial frames of reference are non-equivalent. More detailed calculations using Maxwell's equations lead to the same conclusion. That is, Maxwell's equations are not invariant under Galilean transformation. Therefore, classical electromagnetic theory and the Galilean transformation taken together imply the existence of a preferred frame of reference in which Maxwell's equations are valid. The idea of a preferred frame of reference is foreign to mechanics. A number of attempts were proposed to resolve this conflict. These attempts included the following hypotheses:

(1) Maxwell's equations are presumably wrong and need to be modified to obey the principle of Newtonian relativity;

(2) Presumably there is indeed a preferred frame of reference, that of a stationary luminiferous ether. Maxwell's equations require modification in other inertial frames of reference.

(3) Maxwell's equations are presumed correct and have the same

form in all inertial reference frames. There is some other transformation than the Galilean transformation which will make both electromagnetical and mechanical equations transform in an invariant way.

As we know now, the third proposal is the correct one, but it was not accepted without resistance. The first attempt was abandoned rather quickly. When it was tried, new terms that make Maxwell's equations Galilean invariant led to predictions of new phenomena which did not exist when tested experimentally. So the attempt was abandoned. The second attempt was ruled out after extensive experiment. Towards the latter half of the 19th century physicists were seriously interested in a search for the existence of the ether. Owing to the works of Young and Fresnel, light was viewed as a mechanical wave and its propagation, therefore, requires a physical medium, the ether. Maxwell accepted this ether hypothesis. As light can travel through space, it was assumed that ether must fill all of space and the speed of light must be measured with respect to the stationary ether.

The existence of such an ether would imply a preferred inertial frame of reference - the one at rest relative to the ether. In all other initial frames which move with a constant velocity relative to the ether, measurement and formulation of physical laws would mix both the ether, and the effect of motion relative to the ether. Accordingly, the physical laws would appear different in different inertial frames. Only the preferred frame of reference would reveal the true nature of the physical laws.

The putative existence of ether and the law of velocity addition (according to the Galilean transformation) suggest that it should be possible to detect some variation of the speed of light as emitted by some terrestrial source. As earth travels through space at 30 km/sec in an almost circular orbit around the sun, it is bound to have some relative velocity with respect to the ether. If this relative velocity is added to that of the light emitted from source, then light emitted simultaneously in two perpendicular directions should travel at different speeds, corresponding to the two relative velocities of the light with respect to the ether. Michelson and Morley set out in 1887 to detect this variation in the velocity of the propagation of light. Their ingenious way of doing so depends on the phenomenon of interference of light to determine whether the time taken for light to pass over two equal paths at right angles was different or not.

The ether must be assumed to have contradictory mechanical properties: it must be the softest and also the hardest substance in the world. It has to be assumed softest because all material bodies can pass through it without any resistance from the ether itself. (Otherwise, for example, earth would have slowed down and fallen into the sun during the billions of years of its travelling around the sun.) On the other hand, the ether must be harder than any material because light (ether vibration) travels with such a high speed that its elastic constant must be highest of all known materials. Such contradictions did not prevent physicists of the 19th century from clinging to their belief in this hypothetical ether.

2-2 The Michelson - Morley experiment

The Michelson - Morley experiment was performed with an interferometer, shown in Figure 1. The interferometer is basically comprised

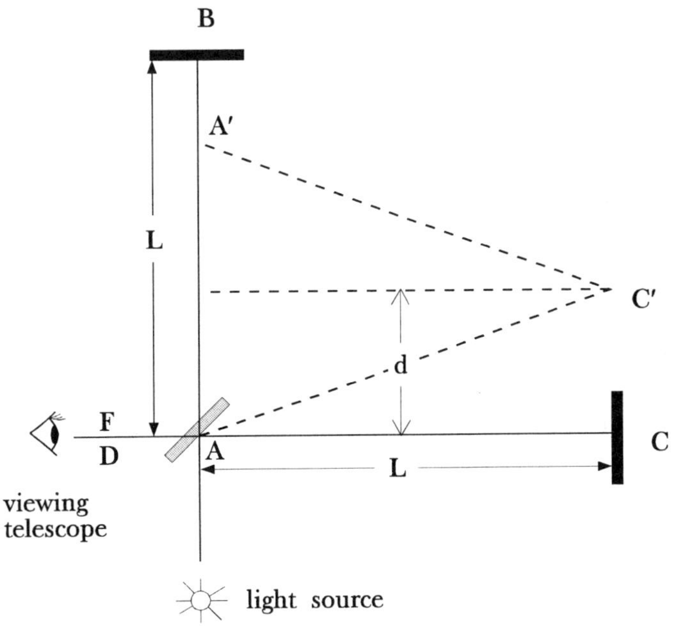

Figure 1: The Michelson Interferometer.

of a light source S, a half-silvered glass plate A, and two mirrors B and C, all mounted on a rigid base. The two mirrors are placed at equal distance L from the plate A. Light from S enters A and splits into two beams. One goes to mirror B which reflects it back. The other goes to mirror C, also to be reflected back. On arriving back at A, the two reflected beams are recombined as two superposed beams, D and F as indicated. If the time taken for light to go from A to B and back is the same as the time from A to C and back, the two beams D and F will be in phase and reinforce each other, but if the two times differ slightly, the two beams will be slightly out of phase and interference will result. We now calculate the two times to see whether they are the same or not. Let us first calculate the time required for the light to go from A to B and back. If the line AB is parallel to the earth's motion in its orbit (at certain times of the day and night), and if the earth is moving at a speed u and the speed of light in the ether is c, then the time is

$$t_1 = L/(c - u) + L/(c + u) = 2L/c[1 - (u/c)^2] \qquad (1)$$

where (c - u) is the upstream speed of light with respect to the apparatus, and (c + u) is the downstream speed.

Our next calculation will be of the time t_2 for the light to go from A to C. We note that while the light goes from A to C, the mirror C moves to the right relative to the ether through a distance $d = ut_2$, to the position C′; in the same time, the light travels a distance ct_2 along AC′. For this right triangle we have

$$(ct_2)^2 = L^2 + (ut_2)^2$$

from which we get

$$t_2 = L/[c^2 - u^2]^{1/2} .$$

Similarly, while the light is returning to the half-silvered plate, the plate moves to the right, to the position A′. The total path length for the return trip is the same, as can be seen from the symmetry of Figure 1. Therefore the return time is also the same, the total time for light to go from A to C and back is then $2t_2$, and we denote it by t_3:

$$t_3 = 2L/[c^2 - u^2]^{1/2} = 2L/c[1 - (u/c)^2]^{1/2}\ . \qquad (2)$$

In Eqs. (1) and (2) the numerators are the same and represent the time that would be taken if the apparatus were at rest relative to the ether. The denominators represent the modifications in the times caused by the motion of the apparatus. As $u \ll c$, we can expand the denominators by using the binomial theorem and obtain

$$t_1 = (2L/c)[1 + u^2/c^2 + ...] \qquad (3)$$

$$t_3 = (2L/c)[1 + u^2/2c^2 + ...] \qquad (4)$$

and the time difference Δt is

$$\Delta t = t_1 - t_2 = -(L/c)(u^2/c^2) = -\ (L/c)\ \beta^2 \qquad (5)$$

where $\beta = u/c$.

It is most likely that we shall be unable to make the two lengths L exactly equal. If so we simply rotate the apparatus 90 degrees, so that AB is perpendicular to the motion and AC is in the line of motion. Then any small difference in length becomes unimportant, and what we look for is a shift in the interference fringes when we rotate the apparatus. Now if the apparatus is rotated through 90 degrees, the difference in the transit times is

$$\Delta t' = t_1' - t_3' = (L/c)\ \beta^2 \qquad (6)$$

which is not equal to Δt.

This difference yields a shift in the interference pattern across the crosshairs of the viewing telescope. If the optical path difference between the beams changes by one wavelength, for example, there will be a shift of one fringe. If δ represent the number of fringes moving past the crosshairs as the pattern shifts, then

$$\delta = (\Delta t - \Delta t')/T = 2L\ \beta^2/\lambda \qquad (7)$$

where $T = \lambda/c$ and λ is the wavelength of the light. So by rotating the apparatus, we should be able to observe a steadily changing fringe shift, with maximum and minimum indicating the direction of the earth's velocity through the ether. And from the magnitude of the fringe shift, we should be able to calculate the value of u by using Eq.(7). In the Michelson and Morley experiment of 1887, the effective length L was 11 m, and sodium light of $\lambda = 5900$ A was used. The orbit velocity of the earth is 3×10^4 m/sec, so $\beta = 10^{-4}$. The expected shift, from Eq.(7), would be about as follows.

$$\delta = (22 \text{ m})(10^{-4})^2/5.9 \times 10^{-5} = 0.37$$

or a shift of about four-tenths of a fringe.

A shift of as little as 0.005 fringe could be detected by the Michelson-Morley interferometer. However, no fringe shift in the interference pattern was observed.

The null result of the Michelson-Morley experiment was very puzzling, and most disturbing at the time. It seemed as if the ether were dragged along by the earth. But if this were so, there would be no stellar aberration at all. Apart from this and other difficulties, its status as an absolute reference frame was gone forever. Many later attempts to save the theory of ether also failed (see *Basic Concepts in Relativity and Early Quantum Theory*, by Resnick and Halliday, Wiley, 1985).

2-3 The postulates of special theory of relativity

Now we are forced by experiment to conclude that an ether hypothesis is untenable. Further, Maxwell's equations are correct and do not need modification. At this point we had best turn our attention to the work of Poincaré. As early as 1899, and again in 1900 and in 1904, Poincaré suggested that the null result of Michelson-Morley experiment was a manifestation of a general principle that absolute motion cannot be detected by laboratory experiments of any kind, and this implied that the laws of nature must be the same in all inertial frames. He called this the principle of relativity. Now as Maxwell's equations involve the velocity of light, if it is to be the same in all inertial frames of reference we need to revise radically the concept of relative velocity as applied to light.

Lorentz, working on these ideas, came very close to the formulation of special relativity, but it was not until 1905 that Einstein realized the full implications and took the key step involved. In his epoch-making paper, "On the Electrodynamics of Moving Bodies" (*Annalen der Physik*, 17(1905). For English translation, see *The Principle of Relativity*, Dover, New York, 1952), he developed the Special Theory of Relativity from two basic postulates which are rephrased as follows:

(1) *The laws of physics are the same in all inertial frames. No preferred inertial frame exists. (The principle of relativity.)*

(2) *The velocity of light in empty space is the same in all inertial frames and is independent of the motion of the emitting body. (The principle of the constancy of the velocity of light.)*

These postulates is often called **the principle of relativity of Einstein**, in contrast to the principle of Newtonian relativity which is based on an infinite speed of propagation of interaction. The implications of these postulates are very wide. They radically revised our concepts of space and time. Absolute space is abolished in its Maxwellian role as an ether which is the carrier of electromagnetic waves. Time is also not absolute any more. Time elapses differently in different frames of reference. As an example, consider two observers S and S′. S′ is moving from stationary S with a constant velocity u. Now S sends a beam of light toward a mirror

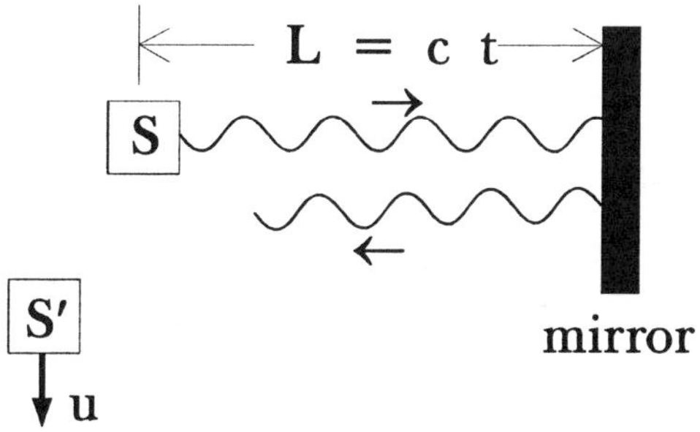

(a) As seen by observer S.

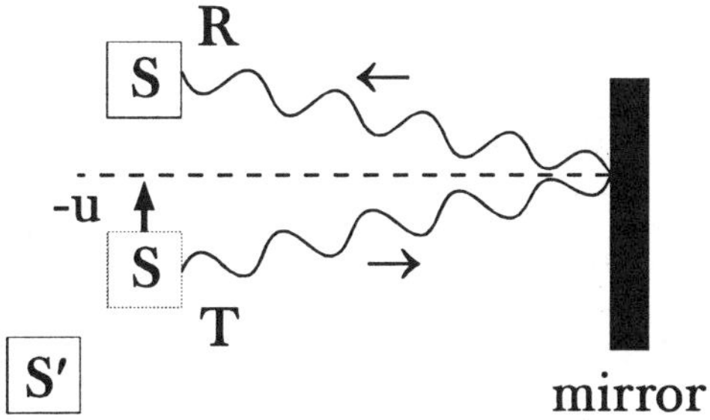

(b) As seen by observer S′.

Figure 2: Time elapses differently in different frames.

at a distance L away and receives the reflected beam back after a time interval $2\Delta t$. To S the beam is sent and received at the same point (Figure 2a). However, if S′ is moving off S in a direction parallel to the mirror (Figure 2b), as seen from S′, the beam is sent out from T and received at R at a time interval $2\Delta t'$ later. During the same time interval S moves, as seen from S′, from T to R through a distance $u \times 2\Delta t$. By an inspection of Figure 2b, we have

$$\overline{TM'} = \overline{M'R} = \frac{\overline{TR}}{2} = u\Delta t'$$

and

$$\overline{TM} = \overline{MR} = \left[L^2 + (u\Delta t')^2\right]^{1/2} \quad .$$

Now the speed of light as measured by S is $L/\Delta t$, while according to S′ the speed of light is given by

$$\frac{\overline{TM}}{\Delta t'} = \left[L^2 + (u\Delta t')^2\right]^{1/2}/\Delta t'$$

According to Einstein's second postulate of the constancy of the speed of light, we have

$$L/\Delta t = [L^2 + (u\Delta t')^2]^{1/2}/\Delta t'$$

or, solving for $\Delta t'$

$$\Delta t' = \Delta t/[1 - (u/c)^2]^{1/2} \ . \qquad (8)$$

Thus the moving observer S′ measures a longer time interval for the same occurrence. This effect is known as time dilation. It is Einstein's second postulate of the constancy of the speed of light that leads to the less familiar results of special relativity. However, for our purposes it will be more efficient to calculate them by use of the Lorentz transformations.

2-4 The Lorentz transformation

As the Galilean transformations are inconsistent with Einstein's postulate of the constancy of the speed of light, we must modify them in such a way that they will incorporate Einstein's postulate and make both mechanical and electromagnetic equations transform in an invariant way. To this end, we consider two inertial frames S and S′. For simplicity, let the corresponding axes of the two frames be parallel and the relative motion is along the x-axis, with frame S′ moving at a constant velocity u relative to S.

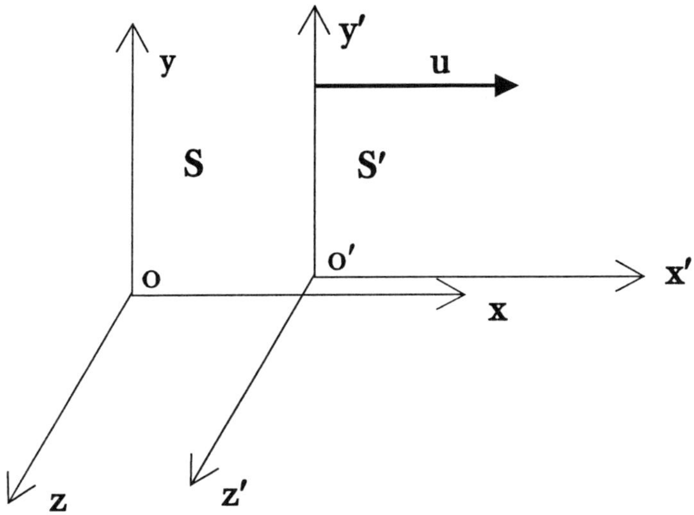

Figure 3: Coordinate frames S and S′ moving with relative speed u.

Define t and t′ to be zero at the instant when the origins 0 and 0′ coincide. Figure 3 represents the viewpoint of observers in S. Since the Galilean transformations are a special case of linear transformations and the Galilean transformations are also a limiting case of our intended transformations, we therefore assume linearity for our intended transformations and consider the transverse dimensions first. Since the relative motion of the coordinate systems occurs only along the x-axis, events that have $y' = 0$ should also have $y = 0$ independently of the values of x', z' or t'. Therefore, if the relation is linear, y can only depend on y', and the relation must be $y = ky'$. The symmetry requires that $y' = ky$, and these two relations can both be true only if $k = 1$. The same argument also applies to the z-axis. Therefore

$$y' = y, \qquad \text{and} \qquad z' = z \,. \tag{9}$$

These relations are the same as in Galilean transformations.

Along the longitudinal dimension, the relation between x and x' must depend on the time, so we consider the most general linear relation

$$x' = ax + bt \,. \tag{10}$$

Now the origin $0'$, where $x' = 0$, corresponds to $x = ut$. Substituting these into Eq.(10), we have

$$0 = a\,ut + b\,t$$

from which

$$b = -\,a\,u$$

and our general linear relation (10) simplifies to

$$x' = a(x - u\,t) \,. \tag{11}$$

By symmetry, we must have

$$x = a(x' + u\,t'). \tag{12}$$

Now we apply Einstein's second postulate of the constancy of the speed of light. If a pulse of light is sent out from the origin 0 of frame S at $t = 0$, its position along the x axis at later time t is given by $x = ct$ and its position along x' axis is $x' = ct'$. Putting these expressions in Eqs. (11) and (12), we obtain

$$ct' = a(c - u)t, \quad \text{and} \quad ct = a(c + u)t'$$

or

$$t/t' = c/a(c - u), \quad \text{and} \quad t/t' = a(c + u)/c \; .$$

Therefore $$c/a(c - u) = a(c + u)/c$$

solving for a,

$$a = 1/[1 - (u/c)^2]^{1/2}$$

Substituting this in Eqs.(11) and (12) gives

$$x' = (x - ut)/[1 - (u/c)^2]^{1/2} \qquad (13)$$

and

$$x = (x' + ut')/[1 - (u/c)^2]^{1/2} \; . \qquad (14)$$

Eliminating either x or x′ from Eqs.(13) and (14), we obtain

$$t' = (t - ux/c^2)/[1 - (u/c)^2]^{1/2} \qquad (15)$$

and

$$t = (t' + ux'/c^2)/[1 - (u/c)^2]^{1/2} \; . \qquad (16)$$

The quantity u/c is usually represented by the Greek symbol β(beta), and $1/[1 - (u/c)^2]^{1/2}$ is called **the Lorentz factor**, usually represented by another Greek symbol γ(gamma).

Combining all of these results, we obtain the Lorentz transformation formulas

$$\begin{array}{ll} x' = \gamma(x - \beta ct) & x = \gamma(x' + \beta ct)' \\ y' = y & y = y' \\ z' = z & z = z' \\ t' = \gamma(t - \beta x/c) & t = \gamma(t' + \beta x'/c) \; . \end{array} \qquad (17)$$

Everything else in Special Relativity can be derived from these relations.

If we let $\beta << 1$, so that $\gamma \simeq 1$, we see that Eqs.(17) reduce to the Galilean transformations that we now see as a first approximation to the Lorentz transformations for $\beta << 1$.

The Lorentz transformations which are based on Einstein's postulates imply a new philosophy of space and time measurements. We now try to understand as much as we can the nature of this new philosophy of observation and derive various consequences. We begin with some definitions. Two events occurring at the same time but not necessarily at the same place are called **simultaneous**; and two events which occur at the same place but not necessarily at the same time are termed **colocal**. Two events which appear at the same place and at the same time are said to be **coincident**. It is clear that when two events are coincident in one inertial frame, then they are coincident in all inertial frames. But we will see soon that simultaneity and colocality no longer have universal meaning. In the following discussions, we still use Figure 3.

A. Relativity of simultaneity; causality

Consider two events in S′ which occur at (x_1', t_1') and (x_2', t_2'). They would appear in our frame S at (x_1, t_1) and (x_2, t_2). According to the Lorentz transformations, we have

$$t_2 - t_1 = \gamma[(t_2' - t_1') + u(x_2' - x_1')/c^2] . \tag{18}$$

It is easy to observe that:

(a) If the two events take place simultaneously in S′, then $t_2' - t_1' = 0$. But the events do not occur simultaneously in S frame for there is a finite time lapse

$$\Delta t = t_2 - t_1 = \gamma u(x_2' - x_1')/c^2 .$$

Moreover, depending on the sign of $(x_2' - x_1')$ the time interval Δt can be positive or negative, i.e. in the frame S the "first" event in S′ can take place earlier or later than the "second" one. The sole exception is the case when two events occur coincidently in S′, in which case they also occur at the same place and at the same time in S.

(b) If the order of the two events in frame S is not reversed in time, then $\Delta t = t_2 - t_1 > 0$ which implies that

$$(t_2' - t_1') + u(x_2' - x_1')/c^2 > 0$$

or

$$(x_2' - x_1')/(t_2' - t_1') < c^2/u$$

and the above will be true as long as

$$(x_2' - x_1')/(t_2' - t_1') < c\ .$$

Thus the order of events will remain unchanged if no signal can be transmitted with a speed greater than c.

The above also illustrates particularly clearly that the theory of relativity is incompatible with the notion of action at a distance.

B. Time dilation; relativity of colocality

Consider two events in S′ which take place at times t_1' and t_2' but at the same place. For simplicity consider this to be on the x′ axis so that $y' = z' = 0$. These two events would appear in our own frame S at (x_1, t_1) and (x_2, t_2). According to the Lorentz transformations, we have

$$\Delta x = u\Delta t'/[1 - u^2/c^2]^{1/2} \tag{19}$$

$$\Delta t = \Delta t'/[1 - u^2/c^2]^{1/2} \tag{20}$$

where $\Delta t' = t_2' - t_1'$ etc. It is easy to observe that:

(a) The two events which occurred at the same place in S′ do not occur at the same place in S. According to Galilean transformations, $\Delta x =$

$u\Delta t'$. The denominator $[1 - u^2/c^2]^{1/2}$ is the relativistic correction to the Galilean result.

(b) In frame S, in which the two events do not occur in the same place ($\Delta x \neq 0$), the time interval Δt becomes longer.

Suppose the events are the ticks of any clock that is at rest in S′. In S, the interval between these ticks is longer: i.e., the moving clock goes more slowly than the one at rest. Time dilation is a fact already confirmed in the laboratory. For example, charged pions have a lifetime $T_0 = 2.6 \times 10^{-8}$ sec when they are at rest. When we observe them in fast motion in a synchrotron, their lifetimes become larger according to

$$T = T_0 /[1 - u^2/c^2]^{1/2} .$$

C. Length contraction

Consider a rod of length L_o in the S′ frame in which the rod lies at rest along the x′ axis: $L_o = x_2' - x_1'$. The rod is moving lengthwise with velocity u relative to the S frame. Now an observer in the S frame makes a simultaneous measurement of the two ends of the rod. The Lorentz transformations give

$$x_1' = \gamma[x_1 - ut_1], \quad x_2' = \gamma[x_2 - ut_2]$$

from which

$$x_2' - x_1' = \gamma[x_2 - x_1] - \gamma u(t_2 - t_1) .$$

Let $L(u) = [x_2 - x_1]$, which is the length of the rod moving with speed u relative to the observer, and remember that $t_1 = t_2$, $\gamma = 1/[1 - u^2/c^2]^{1/2}$, hence the last equation becomes

$$L(u) = [1 - u^2/c^2]^{1/2} L_0 . \tag{21}$$

Thus the length of a moving object, in the direction of motion, is less than the length of the same object at rest by a factor $[1 - u^2/c^2]^{1/2}$.

Does this not violate the symmetry requirement? That is, if all frames of reference are equally valid and if $L' = \gamma L$, doesn't the expression $L = \gamma L'$ have to be equally true? Not so. The measurement was not carried out in the same way in the two frames of reference. The two events (of marking the positions of the two ends of the rod) were simultaneous in the S frame but not simultaneous in the S′ frame. This difference gives the asymmetry of the result. The length L′ was equal to the length L_0 only because the rod was at rest in the S′ frame. As a general expression, $\Delta x' = \gamma \Delta x$ is not true. The full expression relating distances in two frames of reference is $\Delta x' = \gamma(\Delta x - u\Delta t)$. The symmetrical inverse relation is $\Delta x = \gamma(\Delta x' + u\Delta t')$. In the case that was considered above, $t = 0$, so $\Delta x' = \gamma \Delta x$. But $\Delta t' \neq 0$, so Δx did not equal $\Delta x'$.

The dimensions of an object normal to the direction of motion are not affected by the motion. Hence a ball moving at high speed has a shape like that of a discus.

The appearance of a high-speed object is another matter. The act of seeing involves the simultaneous reception of light from the different parts of the object. In order for light from different parts of an object to reach the eye or a camera at the same time, the light from different parts of the object must be emitted at different times, to compensate for the different distances the light must travel. Thus, taking a picture of a moving body, or even looking at it, fails to give a valid impression of its length. Interestingly, the distortion which makes the Lorentz contraction seem to disappear, instead makes the object seem to rotate in space. This was shown for the first time by James Terrell in 1959 (*Physics Review*, 116(1959), 1041; *American Journal of Physics*, 28(1960) 607).

We would like to stress that it is not right to say that motion makes an object **appear** shorter. The moving object **is** shorter, and its appearance is deceptive. The contraction is of a purely kinematical character. No internal tensions causing a deformation arise in the body.

D. Transformations of velocities

It is important to recognize (in spite of the loose language in many books on relativity) that in comparing velocities in different reference frames we are not adding velocities, we are transforming the measurement of a single velocity from one frame of reference to another. If a particle moves from position 1 to position 2 between times t_1 and t_2, the components of velocity are defined as

$$v_1 = (x_2 - x_1)/(t_2 - t_1) = dx/dt,$$

$$v_2 = (y_2 - y_1)/(t_2 - t_1) = dy/dt,$$

$$v_1' = (x'_2 - x'_1)/(t'_2 - t'_1) = dx'/dt',$$

and so on. To work out the transformation of the x-component of velocity, apply Eq.(17) to x_1 and x_2 and take the difference:

$$dx = (dx' + u\ dt')/[1 - u^2/c^2]^{1/2}$$

similarly

$$dx' = (dx - u\ dt)/[1 - u^2/c^2]^{1/2}\ .$$

Do the same for the time intervals:

$$dt = (dt' + udx'/c^2)/[1 - u^2/c^2]^{1/2}$$

$$dt' = (dt - udx/c^2)/[1 - u^2/c^2]^{1/2}$$

and from these we get

$$dx/dt = (dx' + u\ dt')/(dt' + ux'/c^2).$$

Divide both numerator and denominator of the right side of this equation by dt', and remember that $dx'/dt' = v_1'$. The result is

$$v_1 = (v_1' + u)/[1 + uv_1'/c^2]\ . \qquad (22a)$$

This is the transformation equation for the x-component of velocity. Similarly, we can find the y and z components:

$$v_2 = v_2'(1 - u^2/c^2)^{1/2}/[1 + uv_1'/c^2] \tag{22b}$$

$$v_3 = v_3'(1 - u^2/c^2)^{1/2}/[1 + uv_1'/c^2] \; . \tag{22c}$$

It is easy to observe that the transverse velocity components depend on the x-component. Moreover, according to the velocity transformation formulas Eq.(22) the speed of light is the limiting velocity. This can be easily illustrated with an extreme example. Suppose that $u = c$, while $v_1' = c$ also, then Eq.(22a) gives

$$v_1 = (c + c)/(1 + c^2/c^2) = c,$$

and the Galilean transformation would yield $v_1 = v_1' + u = 2c$.

The velocity transformation formulas Eq.(22) also show that the angle is relative and changes in transition from one frame to another. To this end, consider a particle in the S frame moving in the xy plane with the velocity v which makes an angle θ with the x-axis. Then

$$\tan\theta = v_1/v_2; \quad v_1 = v\cos\theta, \quad v_2 = v\cos\theta \; .$$

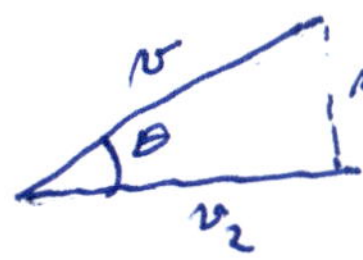

In the S′ frame, we get from Eq.(22)

$$\tan\theta' = v_1'/v_2'$$

$$= v \sin\theta(1 - u^2/c^2)^{1/2}/(v \cos\theta - u) \; . \tag{23}$$

E. Transformation of differential operators

Starting from the Lorentz transformations (17) we can derive a set of transformation formulas of the differentiation operators. To show this, we consider a scalar function 0

$$\phi(x,y,z,t) = \phi(x',y',z't')$$

which can represent, for example, a wave travelling in space in the x direction. Then we can write $\partial\phi/\partial x$ as

$$\partial\phi/\partial x = (\partial\phi/\partial x')(\partial x'/\partial x) + (\partial\phi/\partial t')(\partial t'/\partial x) .$$

Now from the Lorentz transformations (17) we have

$$\partial x'/\partial x = \gamma, \quad \text{and} \quad \partial t'/\partial x = -\gamma V/c^2. .$$

Combining this with the last expression for $\partial\phi/\partial x$ we obtain

$$\partial\phi/\partial x = \gamma\big[(\partial\phi/\partial x') - (V/c^2)(\partial\phi/\partial t')\big] . \tag{24a}$$

Similarly,

$$\begin{aligned} \partial\phi/\partial t &= (\partial\phi/\partial x')(\partial x'/\partial t) + (\partial\phi/\partial t')(\partial t'/\partial t) \\ &= \gamma[(\partial\phi/\partial t') - V(\partial\phi/\partial x')] \end{aligned} \tag{24b}$$

Since Eqs.(24a) and (24b) are true for any ϕ, we have the operator equations

$$\partial_x = (\partial_{x'} - Vc^{-2}\partial_{t'})/(1 - V^2/c^2)^{1/2} \tag{25a}$$

$$\partial_t = (\partial_{t'} - V\partial_{x'})/(1 - V^2/c^2)^{1/2} \tag{25b}$$

where $\partial_x = \partial/\partial x$, $\partial_{x'} = \partial/\partial x'$, etc. Along the transverse dimensions, $\partial_y = \partial_{y'}$ and $\partial_z = \partial_{z'}$. These are the transformations of the differentiation operators corresponding to the transformations of variables Eqs.(17), and they are very useful in dealing with field problems or wave propagation.

2-5 Geometry of space-time (four-dimensional space)

For the space-time of physics we now need four coordinates, the time t and the three space coordinates. We put

$$x^0 = ct, \;\; x^1 = x, \;\; x^2 = y, \;\; x^3 = z \tag{26}$$

so that the four coordinates may be written x^μ, and the suffix μ takes on the four values 0, 1, 2, 3 rather than 1, 2, 3, 4 (in this case, $x^4 = ict = ix^0$). The suffix is written in the upper position (i.e., as superscript) in order that we may maintain a "balancing" of the suffixes in all the general equations of the theory. The precise meaning of "balancing" will become clear a little later.

In terms of the four space-time coordinates, the Lorentz transformations now take on a slightly different appearance:

$$\begin{aligned} x'^1 &= \gamma(x^1 + \beta x^0) & x^1 &= \gamma(x'^1 - \beta x'^0) \\ x'^2 &= x^2 & x^2 &= x'^2 \\ x'^3 &= x^3 & x^3 &= x'^3 \\ x'^0 &= \gamma(x^0 - \beta x^1) & x^0 &= \gamma(x'^0 + \beta x'^1). \end{aligned} \qquad (27)$$

Letting the Greek indices run from 0 to 3, this can, in turn, be distilled into a single equation:

$$x'^\mu = \sum_{\nu=0}^{3} a_{\mu\nu} x^\nu, \qquad \mu = 0,1,2,3 \qquad (28)$$

or, in matrix form

$$\begin{vmatrix} x'^0 \\ x'^1 \\ x'^2 \\ x'^3 \end{vmatrix} = \begin{vmatrix} \gamma & -\gamma\beta & 0 & 0 \\ \gamma\beta & \gamma & 0 & 0 \\ 0 & 0 & 1 & 0 \\ 0 & 0 & 0 & 1 \end{vmatrix} \begin{vmatrix} x^0 \\ x^1 \\ x^2 \\ x^3 \end{vmatrix} .$$

It is customary to use Greek indices (μ and ν etc.) to label four-dimensional variables, Latin indices (i, j etc.) to label three-dimensional variables. One advantage of writing things in the format of Eq.(28) is that we can handle in the same format a more general transformation, in which the relative motion is not along a common coordinate axis; the matrix would be more complicated, but the structure of Eq.(28) would remain unchanged.

Eq.(28) reminds us of the orthogonal rotations which is familiar from classical mechanics; this similarity is no accident! The general Lorentz transformations can indeed be interpreted as an orthogonal rotation of coordinate axes in a four-dimensional space with axes x^1, x^2, x^3, and x^0 = ct. This four-dimensional space is called **world space** or **Minkowski space**. (In fact, one can show that the Lorentz transformations described in Eq.(28) correspond to a rotation in the x^0 - x^1 plane.)

A. Intervals

The concept of **event** will be used frequently. An **event** is taken to mean something happening at a certain place and at a certain time. An event is depicted as a point, called **world point**, in the four-dimensional Minkowski space. The history of a body moving along a path in space is summed up by a line in space-time. This is the **world line**. Let us consider again the two frames S and S′ and observe events in them. Let the first event consist of sending out a pulse of light signal from a point (x_1, y_1, z_1) in the S frame at time t_1. The second event consists of the arrival of the signal at point (x_2, y_2, z_2) at time t_2. The light propagates with the speed c in all directions. By the instant t_2, the signal will reach points of a sphere of the radius $c(t_2 - t_1)$. The radius of the sphere is also equal to $[(x_2 - x_1)^2 + (y_2 - y_1)^2 + (z_2 - z_1)^2]^{1/2}$. Thus, we have the following relation between coordinates of the two events in S frame

$$(x_2 - x_1)^2 + (y_2 - y_1)^2 + (z_2 - z_1)^2 - c^2(t_2 - t_1)^2 = 0 \quad . \tag{29}$$

The same two events can be observed from S′ frame. Then, according to Einstein's second postulate of invariance of the velocity of light, we have

$$(x'_2 - x'_1)^2 + (y'_2 - y'_1)^2 + (z'_2 - z'_1)^2 - c^2(t'_2 - t'_1)^2 = 0 \ .$$

The quantity

$$s_{12} = [c^2(t_2 - t_1)^2 - (x_2 - x_1)^2 - (y_2 - y_1)^2 - (z_2 - z_1)^2]^{1/2} \tag{30}$$

is called the **interval** or the **line element** between the two events. It follows from the postulate of invariance of the velocity of light that if the line element between two events is zero in one coordinate frame, then it is equal to zero in all other frames.

If two events are infinitesimally close to each other, the interval ds between them is

$$ds^2 = c^2dt^2 - dx^2 - dy^2 - dz^2$$

or

$$ds^2 = (dx^o)^2 - (dx^1)^2 - (dx^2)^2 - (dx^3)^2 \tag{31}$$

where ds^2 can be regarded, from the formal point of view, as the square of the distance between two world points in Minkowski space. We may write ds^2 in a more general form

$$ds^2 = \sum_{\mu,\nu=0}^{3} g_{\mu\nu} dx^\mu dx^\nu \tag{32}$$

where $g_{\mu\nu}$ is as follows:

$$g_{00} = 1, \quad g_{11} = g_{22} = g_{33} = -1, \quad g_{\mu\nu} = 0, \quad \text{if} \quad \mu \neq \nu \;. \tag{32a}$$

The real usefulness of this notation becomes clear later.

If $ds = 0$ in frame S, then $ds' = 0$ in frame S′. Furthermore, ds and ds′ are infinitesimals of the same order. It follows that ds and ds′ must be proportional to each other,

$$ds^2 = a\, ds'^2 \tag{33}$$

where the proportionality constant a can depend only on the absolute value of the relative velocity of the two inertial frames. Owing to the homogeneity of space and time and the isotropy of space, the coefficient a cannot depend on the coordinates, or the time, or the direction of the relative velocity. Because of the complete equivalence of the two frames S′ and S, we can write on the same grounds that

$$ds'^2 = a\, ds^2.$$

Combining this with Eq.(33), we find that

$$a^2 = 1, \qquad \text{or} \qquad a = \pm 1 \; .$$

The sign of the interval in all frames must be the same. We take the positive sign, and so the value of a equal to -1 is discarded. We thus arrive at the conclusion that

$$ds^2 = ds'^2.$$

From the equality of the infinitesimal intervals, there follows the equality of finite intervals,

$$s'^2 = s^2.$$

Thus we arrive at a very important result: the interval between two events is the same in all inertial frames of reference. This invariance of the interval between two events is the mathematical expression of the constancy of the velocity of light.

The invariance of the interval ds^2 is a powerful tool in many of its applications. The skilful use of this invariance often avoids an explicit Lorentz transformation. Insight into the nature of the interval is gained by considering some special cases. First, we introduce the notation

$$t_{12} = t_2 - t_1, \qquad d_{12} = (x_2 - x_1)^2 - (y_2 - y_1)^2 - (z_2 - z_1)^2 \; .$$

The interval between two events in the S frame can be written as

$$s_{12}{}^2 = c^2 t_{12}{}^2 - d_{12}{}^2. \tag{34}$$

If the two events occur at the same point in S′ frame, then $d'_{12} = 0$, and we have, because of the invariance of intervals,

$$s_{12}{}^2 = c^2 t_{12}{}^2 - d_{12}{}^2 = c^2 t'_{12}{}^2 > 0 \tag{35}$$

in which the interval is real. Real intervals are called **time-like**. The time interval between events in S′ frame is

$$t'_{12} = s_{12}/c = [c^2 t_{12}^2 - d_{12}^2]^{1/2}/c \ .$$

Every time-like interval that connects event 1 with another event lies within the light cones bounded by x = +ct in Figure 4. All events that could have affected event 1 lie in the past light cone; all events that event 1 is able to affect lie in the future light cone. For simplicity, in Figure 4 we plot one space axis (the x axis) versus time. If we include a y axis coming out of the page, the "wedges" become cones — and, with an undrawable z axis, hypercones. Because their boundaries are the trajectories of light rays (no material object can travel faster than light), we call them light cones.

On the other hand, if the two events occur at one and the same time (simultaneously) in the S′ frame, then $t'_{12} = 0$, we have

$$s_{12}^2 = c^2 t_{12}^2 - d_{12}^2 = - d'^2_{12} < 0 \qquad (36)$$

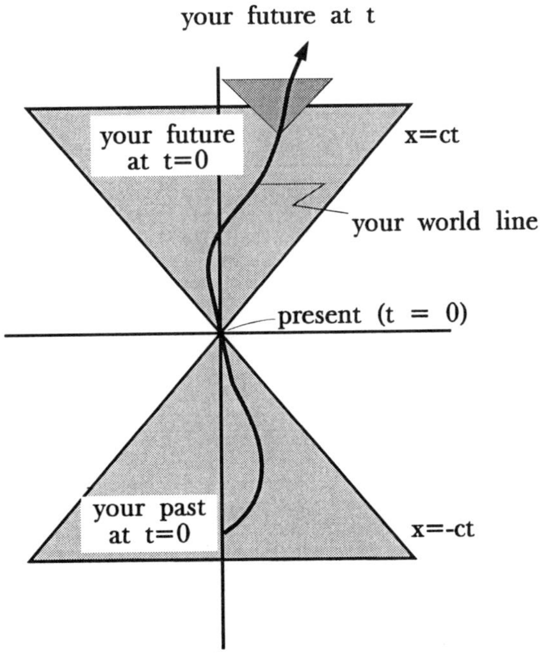

Figure 4: The light cone.

and the interval is imaginary. Such an interval is called **space-like**. There is no causal relationship between events 1 and 2.

Every event that is connected with event 1 by a space-like interval lies outside the light cone of event 1, and it has neither interacted with event 1 in the past, nor is it capable of interacting with it in the future.

When two events can be connected with a light signal only, then

$$s_{12} = 0 \tag{37}$$

and such an interval is said to be **light-like**. Events that can be connected with event 1 by light-like intervals lie on the boundaries of the light cones.

The world line of a particle (i.e., the path of a particle in space-time) must lie within its light cones.

The division of intervals into light-like, space-like, and time-like intervals is, because of their invariance, an absolute concept. This means that the time-like, space-like, or light-like character of an interval is independent of the reference frame.

B. Proper time

The concept of proper time will be very useful in our later discussion. To introduce this concept, consider an object moving uniformly at a velocity u relative to our rest frame S. Assume that two events occur in the frame S′ (relative to which the object is at rest) and are separated by a time interval dt (as read by a clock in S frame). In the frame S′, the time interval between the two events is

$$dt' = ds/c$$

while $dx'^1 = dx'^2 = dx'^3 = 0$. It is clear that the time interval dt' is measured by a clock moving together with the object relative to S. The time measured by means of a clock moving together with a body is called the **proper time** of the body, and it is customarily denoted by the symbol τ, so we now have

$$d\tau = ds/c \, . \qquad (38)$$

Since ds is an invariant and c is a constant, the proper time $d\tau$ is an invariant.

Let us now find the relation between the proper time $d\tau$ and the time read by a clock in S frame relative to which the object (S′ frame) moves at a constant velocity u. As seen from S frame, we have

$$ds^2 = c^2dt^2 - (dx^1)^2 - (dx^2)^2 - (dx^3)^2 \, .$$

Because of the invariance of intervals, we have

$$c^2d\tau^2 = c^2dt^2 - [(dx^1)^2 + (dx^2)^2 + (dx^3)^2]$$

from which

$$d\tau = dt\{1 - [(dx^1)^2 + (dx^2)^2 + (dx^3)^2]/c^2dt^2\}^{1/2}.$$

But

$$[(dx^1)^2 + (dx^2)^2 + (dx^3)^2]/dt^2 = u^2$$

where u is the velocity of the moving clock relative to S frame; and so

$$d\tau = ds/c = dt\{1 - u^2/c^2\}^{1/2} \, . \qquad (39)$$

Eq.(39) is the expression of the time dilation. It indicates that a moving clock goes more slowly than the one at rest.

C. Four-vectors

In an ordinary three-dimensional space, we can define scalar and vector quantities in terms of transformation properties. A vector is a set of three quantities which transform under a change of coordinates like the components of the radius vector. A scalar is a quantity which is unaffected by coordinate transformations.

By analogy with the three-dimensional case, the coordinates of an event (x^0, x^1, x^2, x^3) can be considered as the components of a four-dimensional radius vector (or a radius four-vector) in a four-dimensional Minkowski space. The square of the "length" of the radius four-vector is given by

$$(x^0)^2 - (x^1)^2 - (x^2)^2 - (x^3)^2 \, , \qquad (40)$$

and it does not change under Lorentz transformations (or under any rotations of the four-dimensional coordinate system.)

In general, any set of four quantities $A^\mu (\mu = 0, 1, 2, 3)$ that transform like the components of the radius four-vector x^μ under transformations of the four-dimensional coordinate system is called a **contravariant four-vector**. Under Lorentz transformations

$$\left.\begin{aligned} A^0 &= \gamma(A'^0 + \beta A'^1), & A^1 &= \gamma(A'^1 + \beta A'^0) \\ A^2 &= A'^2, & A^3 &= A'^3 \end{aligned}\right\} \qquad (41)$$

The square length of any four-vector is defined analogously to the radius four-vector,

$$(A^0)^2 - (A^1)^2 - (A^2)^2 - A^3)^2 \, . \qquad (42)$$

With a second contravariant vector B^μ, we have the scalar product invariant

$$A^0B^0 - A^1B^1 - A^2B^2 - A^3B^3 \qquad (43)$$

In order to simplify the writing of such invariants we introduce the device of lowering suffixes. Define

$$A_0 = A^0, \;\; A_1 = -A^1, \;\; A_2 = -A^2, \;\; A_3 = -A^3 \qquad (44)$$

or, using $g_{\mu\nu}$

$$A_\mu = g_{\mu\nu} A^\nu. \qquad (44a)$$

The $g_{\mu\nu}$ is given by Eq.(32a). The quantities A_μ are called the **covariant** components of four vectors. With the two types of four vectors, the expression of (43) may be written

$$\sum_{\mu=0}^{3} A_\mu A^\mu = A_\mu A^\mu . \tag{43a}$$

We can eliminate the summation sign by introducing **Einstein's summation convention:** the repeated indexes appearing once in the lower and once in the upper position are automatically summed over. It is important to remember that the indexes repeated in the lower part or upper part alone are not summed over. For example, $g_{\mu\mu} = 1$, and $x_i x_i = x_i^2$, but $x_\mu x^\mu = (x^0)^2 - (x^1)^2 - (x^2)^2 - (x^3)^2$, and

$$\begin{aligned} A_\mu A^\mu &= A_0 A^0 + A_1 A^1 + A_2 A^2 + A_3 A^3 \\ &= (A^0)^2 - (A^1)^2 - (A^2)^2 - (A^3)^2 \quad . \end{aligned}$$

The $g_{\mu\nu}$ is a device to lower the indexes. Likewise, we can define a $g^{\mu\nu}$ to raise indexes. In the Cartesian coordinates used here

$$g^{\mu\nu} = g_{\mu\nu}, \qquad \text{and} \qquad g_{\mu\nu} g^{\mu\sigma} = \delta^\sigma_\nu$$

D. Four-velocity and four-acceleration

The position vector of a point in Minkowski space is a four-vector x^μ (Note: We denote a four-vector by a representative component, thus x^μ). We now define four-vectors of velocity and acceleration. It is evident that the set of the four quantities dx^μ/dt does not have the properties of a four-vector because dt is not an invariant and therefore $(dx^\mu/dt)(dx^\mu/dt)$ does not retain its value under Lorentz transformations. However the proper time $d\tau (= ds/c)$ is an invariant. Observers in different frames may disagree about the time interval between events (because each is using his own time axis) but all agree on the value of the time interval which would be observed in the frame moving with the particle. The components of the four-velocity of a particle are therefore defined as

$$u^\mu = dx^\mu/d\tau = c\,dx^\mu/ds . \tag{45}$$

With the help of Eq.(39), we may rewrite Eq.(45) completely in terms of quantities observed in frame S as

$$u^{\mu} = [1 - \beta^2]^{-1/2} dx^{\mu}/ds \ . \qquad (46)$$

In terms of the ordinary velocity components v_1, v_2, v_3 we have

$$u^{\mu} = \left(c\left[1 - \beta^2\right]^{-1/2},\ \vec{v}\left[1 - \beta^2\right]^{-1/2}\right) .$$

The four-velocity has now been constructed so that the components are covariant under Lorentz transformations, and its length must therefore be invariant, just as the length of any three-dimensional vector is invariant under rotation. We can easily demonstrate that this is so, for

$$(u^{\mu})^2 = -v^2\left[1 - \beta^2\right]^{-1} + c^2\left[1 - \beta^2\right]^{-1} = +c^2 \ . \qquad (47)$$

Similarly, a four-vector with the components

$$w^{\mu} = d^2x^{\mu}/d\tau^2 = du^{\mu}/d\tau = c^2 du^{\mu}/ds \qquad (48)$$

is called the four-dimensional acceleration of a particle. Differentiate Eq.(47) with respect to τ, and we obtain

$$w^{\mu}u^{\mu} = 0 \qquad (49)$$

i.e. the four-vectors of velocity and acceleration are "mutually perpendicular".

E. Four-momentum vector

We now generalize the definitions of momentum and energy so that in the absence of external forces the momentum and energy of a system of particles are conserved. In Newtonian mechanics the momentum $\vec{p}$ of a particle is defined as $m\vec{v}$, the product of a particle's inertial mass and its velocity. A plausible generalization of this definition is to use the four-velocity u^{μ} and an invariant scalar m_0 which truly characterize the inertial mass of the particle and define the momentum four-vector (four-momentum, for short) P^μ as

$$P^\mu = m_0 u^\mu \tag{50}$$

where the mass of the particle must be its proper mass, i.e., the mass must be that measured in the frame of reference in which the particle is at rest. We customarily call this mass the rest mass of the particle, and denote it by m_0. We can write the four-momentum in component form by using Eq.(46):

$$P_j = m_0 v_j/(1 - \beta^2)^{1/2}, \quad j = 1,2,3$$

and (51)

$$P_0 = m_0 c/(1 - \beta^2)^{1/2} \ .$$

We see that as $\beta = v/c \rightarrow 0$, the first three components of the four-momentum P^μ reduce to $m_0\vec{v}_j$, the components of the ordinary momentum. We can rewrite Eq.(51) as

$$\vec{P}_j = m\vec{v}_j, \quad j = 1,2,3 \tag{52}$$

where

$$m = m(v) = m_0/(1 - \beta^2)^{1/2} \ . \tag{53}$$

Thus, if we wish to interpret the momentum of a particle in the old sense, the "mass" m is no longer an invariant but depends upon the velocity in the particular reference frame. The mass m is called the relativistic mass of the particle. The time component of P_0 can be written as

$$P_0 = m_0 c/(1 - \beta^2)^{1/2} = E/c. \tag{54}$$

As we will see later, the quantity E is the total energy of the particle. For low velocities, the quantity E reduces to

$$E = m_0 c^2 (1 - \beta^2)^{-1/2} = m_0 c^2 + (1/2) m_0 u^2.$$

The second term on the right is the ordinary kinetic energy of the particle, and the first term can be interpreted as the rest mass energy of the particle. Accordingly, we may call E the total energy of the particle. We now write the four-momentum as

$$P^{\mu} = \left(E/c,\ \vec{P}j\right) . \tag{55}$$

The four-momentum has been so constructed that the components are covariant under Lorentz transformations, and its length must therefore be invariant, just as the length of the four-velocity is invariant under Lorentz transformations. We can demonstrate that this is so:

$$P^{\mu}P^{\mu} = (m_0 u^{\mu})(m_0 u^{\mu}) = m_0^2 c^2.$$

However, Eq.(55) gives

$$P^{\mu}P^{\mu} = E^2/c^2 - P^2 .$$

Combining these two, we arrive at the relation

$$E^2/c^2 - P^2 = m_0^2 c^2. \tag{56}$$

The relativistic energy E and the momentum P of a moving object are different when measured with respect to different reference frames. But the combination "E^2/c^2 - P^2" has the same value for all frames of reference, namely $m_{02}c^2$. This relation is very useful.

2-6 Vectors and tensors

We introduced the term "four-vector" for any quantity, such as dx^{μ} or p^{μ}, which undergoes the transformation

$$A^{\mu} \rightarrow A'^{\mu} = a_{\mu\nu}A^{\nu} \tag{57}$$

when there is coordinate transformation

$$x^{\mu} \rightarrow x'^{\mu} = a_{\mu\nu} x^{\nu} \tag{58}$$

where $a_{\mu\nu}$ is the Lorentz transformation matrix given by Eq.(28), and satisfies the orthogonality conditions

$$\sum_{\lambda=0}^{3} a_{\lambda\mu} a_{\lambda\nu} = \delta_{\mu\nu}, \qquad \mu,\nu = 0,1,2,3 \tag{59}$$

which follows directly from the invariance of intervals $s'^2 = s^2$ or $\Sigma(x'^{\mu})^2 = \Sigma(x^{\mu})^2$. More precisely, A^{μ} should be called a contravariant four-vector, to distinguish it from a covariant four-vector defined as a quantity A_{μ} which obeys the transformation rule

$$A_{\mu} \rightarrow A'_{\mu} = a_{\mu\nu} A_{\nu} \,. \tag{60}$$

In metric space, they can be interchanged with the help of $g_{\mu\nu}$:

$$A_{\mu} = g_{\mu\nu} A^{\nu} \,. \tag{61}$$

Since the determinant g does not vanish, the equations can be solved for A^{ν} in terms of the A_{μ}. The result is

$$A^{\nu} = g^{\mu\nu} A_{\mu} \,. \tag{62}$$

Each $g^{\mu\nu}$ equals the cofactors of the corresponding $g_{\mu\nu}$ in the determinant of the $g_{\mu\nu}$, divided by the determinant itself. And

$$g_{\mu\nu} g^{\nu\rho} = \delta^{\rho}_{\mu} \tag{63}$$

where δ^{ρ}_{μ} is the kronecker delta symbol: $\delta^{\rho}_{\mu} = 1$ if $\mu = \rho$, and $\delta^{\rho}_{\mu} = 0$ if $\mu \neq \rho$.

For every contravariant four-vector A^{μ} there corresponds a covariant four-vector A_{μ}, and vice versa.

Although any vector can be written in a contravariant or a covariant form, there are some vectors, such as dx^{μ}, which appear more naturally contravariant and some which appear more naturally covariant. An example of the latter is the gradient $\partial/\partial x^\mu$, which obeys the transformation rule

$$\frac{\partial}{\partial x^\mu} \rightarrow \frac{\partial}{\partial x'^\mu} = \frac{\partial x^\nu}{\partial x'^\mu}\frac{\partial}{\partial x^\nu} .$$

The inverse transformation of Eq.(58) is

$$x^\mu = \tilde{a}_{\mu\nu}x'^\nu = a_{\nu\mu}x'^\nu$$

from which it follows that

$$\frac{\partial x^\nu}{\partial x'^\mu} = a_{\mu\nu} .$$

and therefore the gradient is covariant:

$$\frac{\partial}{\partial x'^\mu} = a_{\mu\nu}\frac{\partial}{\partial x^\nu} . \tag{64}$$

The scalar (inner) product of a contravariant with a covariant four-vector is invariant, that is

$$A'_\mu B'^\mu = \sum_{\mu=0}^{3} a_{\lambda\mu}a_{\mu\beta}A_\lambda B^\beta = A_\lambda A^\lambda . \tag{65}$$

Since the gradient is a covariant, the divergence of a contravariant vector $\partial A^\mu/\partial x^\mu$ is invariant. The d'Alembertian operator is also invariant.

$$\Box^2 = \delta_{\mu\nu}\frac{\partial}{\partial x^\mu}\frac{\partial}{\partial x^\nu} = \nabla^2 - \frac{\partial^2}{\partial t^2} . \tag{66}$$

From two contravariant four-vectors, say A^μ and B^μ, we may form the sixteen quantities $A^\mu B^\mu$. These sixteen quantities form the components of a new quantity, called the **tensor**. It is sometimes called the outer (or direct) product of the vectors A^μ and B^μ, as distinct from the inner (scalar) product.

More precisely, the tensor $A^\mu B^\mu$ is a contravariant tensor of the second rank, and we denote it by $T^{\mu\nu}$:

$$T'^{\mu\nu} = a_{\mu\alpha}a_{\nu\beta}T^{\alpha\beta} . \tag{67}$$

We may multiply more than two contravariant vectors to form contravariant tensors of higher rank. The total number of the indexes of a tensor is called its rank (or order). Similarly, we may form covariant tensors from covariant vectors. Vectors are often called first order tensors, and scalars are called zeroth order tensors.

The definition of Lorentz transformations tells us immediately that $g_{\mu\nu}$ is a covariant tensor of second rank,

$$g_{\mu\nu} = a_{\mu\alpha} a_{\nu\alpha} g_{\alpha\beta} \ .$$

We may use this fundamental tensor to lower or raise any indexes of a tensor, for example

$$A_\mu = g_{\mu\nu} A^\nu, \qquad A^\mu = g^{\mu\nu} A_\nu \ .$$

We may set a subscript equal to superscript in a tensor. We then have to sum over all values of this index, and we are left with a tensor having two fewer free indices than the original one. This process is called **contraction**. For example, if we start with a fourth-order tensor $T^{\mu}{}_{\nu\rho}{}^{\sigma}$, we may set $\sigma = \rho$ and we are left with a second-order tensor $T^{\mu}{}_{\nu\rho}{}^{\rho}$. We may contract again (setting $\mu = \nu$) to get the scalar $T^{\mu}{}_{\mu\rho}{}^{\rho}$.

Since the differentiation $\partial_\mu = \dfrac{\partial}{\partial x^\mu}$ behaves as a vector, we can obtain new tensors by differentiating tensors; for example, if $T^{\beta\gamma}$ is a tensor then

$$T_\alpha^{\beta\gamma} = \frac{\partial}{\partial x^\alpha} T^{\beta\gamma}$$

is a tensor:

$$T'^{\beta\gamma}_\alpha = \frac{\partial}{\partial x'^\alpha} T'^{\beta\gamma} = a_{\alpha\delta} \frac{\partial}{\partial x^\delta} a_{\beta\epsilon} a_{\gamma\zeta} T^{\epsilon\zeta} = a_{\alpha\delta} a_{\beta\epsilon} a_{\gamma\zeta} T^{\epsilon\zeta}$$

which shows $T'^{\beta\gamma}_\alpha$ is also a tensor.

Under certain conditions, new tensors can also be formed by integration. To show this, consider the differentials

$$d\Omega = dx^0 dx^1 dx^2 dx^3 = c\,dt\,dV$$

$$dx^\mu = (dx^0,\ dx^1,\ dx^2,\ dx^3)$$

$$dS_\mu = (dx^2 dx^3 dx^0,\ dx^1 dx^3 dx^0,\ dx^1 dx^2 dx^0,\ dx^1 dx^2 dx^3)$$

$$dS^{\mu\nu} = dx^\mu dx^\nu\ .$$

We can construct the integrated quantities, for example

$$\Lambda = \int \Phi d\Omega \quad \text{scalar,} \qquad \Lambda = \int A_\mu dx^\mu \quad \text{scalar}$$
$$A^\mu = \int T^{\mu\nu} dS_\nu \quad \text{vector,} \qquad \Lambda = \int U^\mu dS_\mu \quad \text{scalar, etc.}$$

Among these expressions, the following are important:

$\oint A_\mu dx^\mu$	line integration,
$\int dx^\mu dx^\nu B_{\mu\nu}$	two-surface integration
$\int A^\mu dS_\mu$	three-surface integration
$\int \Phi d\Omega$	space-time integration.

There are theorems that enable us to transform four-dimensional integrals, analogous to the theorems of Gauss and Stokes in three-dimensional vector analysis.

The integral over a closed hypersurface can be transformed into an integral over the four-volume contained within it by replacing the element of integration dS_μ by the operator

$$dS_\mu \rightarrow d\Omega \frac{\partial}{\partial x^\mu}\ . \tag{68}$$

For example, for the integral of a vector A^μ we have

$$\oint A^\mu dS_\mu = \int \frac{\partial A^\mu}{\partial x^\mu} d\Omega\ . \tag{69}$$

This formula is the generalization of Gauss' theorem. Thus, when $\frac{\partial}{\partial x^{\nu}} A^{\nu} = 0$, the result of the integration is a true scalar and is independent of the choice of the three-surface. Similarly, if $\frac{\partial}{\partial x^{\nu}} T^{\mu\nu} = 0$, then

$$P^{\mu} = \int_{s} T^{\mu\nu} dS_{\nu} \tag{70}$$

is a vector independent of the surface of integration. These statements will be useful later.

2-7 The energy-momentum tensor

We can form a four-momentum vector from the energy and the three-dimensional momentum. We now generalize it in tensor form. Having in mind later applications to the gravitational field and also to simplify the calculation for now, we present the derivation in a general form, not specifying the nature of the system. Consider any system whose action integral has the form

$$I = \int \mathcal{L}\left(q, \frac{\partial q}{\partial x^{\mu}}\right) dV dt = (1/c) \int \mathcal{L} d\Omega \tag{71}$$

where $\mathcal{L}$ is some function of the quantities q, describing the state of the system, and of their first derivatives with respect to co-ordinates and time. (For brevity we write here only one of the q's.) Note that the space integral $\int \mathcal{L} dV$ is the Lagrangian of the system, so that $\mathcal{L}$ can be considered to be the **Lagrangian density**. If the system is closed, then the Lagrangian density does not depend explicitly on the x^{μ}, comparable to the situation for a closed system in mechanics, where the Lagrangian does not depend explicitly on the time.

For simplicity we will write $q_{,\mu} = \partial q / \partial x^{\mu}$. Now consider the variation of I with respect to q and subject to the end conditions $\delta q(1) = \delta q(2) = 0$:

$$\delta I = \frac{1}{c}\int\left(\frac{\partial\mathcal{L}}{\partial q}\delta q + \frac{\partial\mathcal{L}}{\partial q_{,\mu}}\delta q_{,\mu}\right)d\Omega$$

$$= \frac{1}{c}\int\left[\frac{\partial\mathcal{L}}{\partial q}\delta q + \frac{\partial}{\partial x^{\mu}}\left(\frac{\partial\mathcal{L}}{\partial q_{,\mu}}\delta q\right) - \delta q\frac{\partial}{\partial x^{\mu}}\frac{\partial\mathcal{L}}{\partial q_{,\mu}}\right]d\Omega \qquad (72)$$

$$= 0 \; .$$

The second term in the integrand can be simplified by Gauss' theorem:

$$\int\frac{\partial}{\partial x^{\mu}}\left(\frac{\partial\mathcal{L}}{\partial q_{,\mu}}\delta q\right)d\Omega = \oint\frac{\partial\mathcal{L}}{\partial q_{,\mu}}\delta q\, dS_{i}$$

which vanishes upon integration over all space. Thus Eq.(72) reduces to

$$\delta I = \frac{1}{c}\int\left[\frac{\partial\mathcal{L}}{\partial q} - \frac{\partial}{\partial x^{\mu}}\frac{\partial\mathcal{L}}{\partial q_{,\mu}}\right]\delta q\, d\Omega = 0 \; .$$

Since δq is arbitrary except $\delta q(1) = \delta q(2) = 0$, it follows that a necessary condition for $\delta I = 0$ is that the square bracket vanishes, yielding the following equations

$$\frac{\partial}{\partial x^{\mu}}\frac{\partial\mathcal{L}}{\partial q_{,\mu}} - \frac{\partial\mathcal{L}}{\partial q} = 0 \; . \qquad (73)$$

(Don't forget to sum over repeated indexes). From Eq.(73) we can introduce some conserved quantities. In this respect, we first write

$$\frac{\partial\mathcal{L}}{\partial x^{\mu}} = \frac{\partial\mathcal{L}}{\partial q}\frac{\partial q}{\partial x^{\mu}} + \frac{\partial\mathcal{L}}{\partial q_{,\nu}}\frac{\partial q_{,\nu}}{\partial x^{\mu}} \; .$$

Substituting Eq.(73) and noting that $q_{,\mu,\nu} = q_{,\nu,\mu}$, we find

$$\frac{\partial\mathcal{L}}{\partial x^{\mu}} = \frac{\partial}{\partial x^{\nu}}\left(\frac{\partial\mathcal{L}}{\partial q_{,\nu}}\right)q_{,\mu} + \frac{\partial\mathcal{L}}{\partial q_{,\nu}}\frac{\partial q_{,\mu}}{\partial x^{\nu}} = \frac{\partial}{\partial x^{\nu}}\left(q_{,\mu}\frac{\partial\mathcal{L}}{\partial q_{,\nu}}\right) \; . \qquad (74)$$

We can also write

$$\frac{\partial \mathcal{L}}{\partial x^{\nu}} = \delta_{\mu}^{\nu} \frac{\partial \mathcal{L}}{\partial x^{\mu}} .$$

Combining this with Eq.(74) gives

$$\frac{\partial}{\partial x^{\nu}} \left(q_{,\mu} \frac{\partial \mathcal{L}}{\partial q_{,\nu}} \right) = \delta_{\mu}^{\nu} \frac{\partial \mathcal{L}}{\partial x^{\nu}}$$

or

$$\frac{\partial}{\partial x^{\nu}} \left[q_{,\mu} \frac{\partial \mathcal{L}}{\partial q_{,\nu}} - \delta_{\mu}^{\nu} \mathcal{L} \right] = 0 . \tag{75}$$

Introducing the notation

$$T_{\mu}^{\nu} = q_{,\mu} \frac{\partial \mathcal{L}}{\partial q_{,\mu}} - \delta_{\mu}^{\nu} \mathcal{L} \tag{76}$$

Eq.(75) becomes

$$\frac{\partial}{\partial x^{\nu}} T_{\mu}^{\nu} = 0 . \tag{77}$$

If there are several quantities q_{α}, then in place of Eq.(76) we have

$$T_{\mu}^{\nu} = \sum_{\alpha} q_{\alpha,\mu} \frac{\partial \mathcal{L}}{\partial q_{\alpha,\nu}} - \delta_{\mu}^{\nu} \mathcal{L} . \tag{76a}$$

It follows from the generalized Gauss' theorem that Eq.(76a) asserts that the vector

$$P^{\mu} = \text{const.} \int T^{\mu\nu} dS_{\nu} \tag{78}$$

is conserved independent of the surface of integration (see Eqs.(69) and (70)). What is the physical meaning of this four-vector? Before we answer this question, let us bear in mind that the above results are very general. They can be applied either to discrete particles or to problems in which matter and energy is continuously distributed in space. The continuous distributions may be considered as a limiting case of discrete particles, wherein each volume element has the number of particles tending to infinity while the mass of each tends to zero. Conversely, a particle may be considered as a limiting case of a continuous distribution where at some points (i.e., the locations of the particles) the densities tend to become infinitely large. With this understanding, we now consider the energy density e(x). We usually write as the total energy $E = \int e(x)dV$, where dV is the volume element. However, it is easy to see that this expression is not quite right, on two counts. First, dV is the fourth component of a four-dimensional surface element $dS_0 = dxdydz$. For relativistic reasons we must make use of all the components dS_μ. Second, the energy is the fourth component of the energy-momentum vector (i.e., the four-momentum vector). It is clear that the general expression for the energy-momentum vector must be a three-surface integration of a tensor which characterizes the physical system. Therefore, we identify P^μ in Eq.(78) as the energy-momentum vector of the system, and we choose the constant factor in front of the integral so that, in accord with our previous definition, the time component P^0 is equal to the energy of the system multiplied by 1/c. To do this we note that

$$P^0 = \text{const.} \int T^{0\mu} dS_\mu = \text{const.} \int T^{00} dV$$

if the integration is extended over the hyperplane $x^0 = $ constant. On the other hand, according to Eq.(76),

$$T^{00} = \dot{q}\frac{\partial \mathcal{L}}{\partial \dot{q}} - \mathcal{L}, \qquad (\dot{q} = dq/dt)$$

and comparing with the usual formulas relating the energy and the Lagrangian, we see that the quantity T^{00} must be considered as the energy density of the system, and so $\int T^{00}dV$ is the total energy of the system. Thus we must set "constant" = -1/c and we finally get, for the four-momentum vector of the system, the expression

$$P^\mu = \frac{1}{c}\int T^{\mu\nu} dS_\nu \;. \tag{79}$$

The tensor $T^{\mu\nu}$ is called the energy-momentum tensor of the system. It is also known as the stress-energy tensor.

The energy-momentum tensor $T_{\mu\nu}$ as defined by Eq.(76) is not unique. Any other tensor of the form

$$T^{\mu\nu} + \frac{\partial}{\partial x^{\nu}}\phi^{\mu\nu\lambda}, \qquad \phi^{\mu\nu\lambda} = -\phi^{\mu\lambda\nu}$$

will also satisfy Eq.(77). To define the tensor $T^{\mu\nu}$ uniquely, we can use the requirement that the four-tensor of angular momentum of the system must be expressed in terms of the four-momentum by

$$M^{\mu\nu} = \int (x^{\mu} dP^{\nu} - x^{\nu} dP^{\mu}) = \frac{1}{c}\int (x^{\mu} T^{\nu\lambda} - x^{\nu} T^{\mu\lambda}) dS_{\lambda} . \qquad (80)$$

In other words, its "density" is expressed in terms of the "density" of momentum by the usual formula.

The law of conservation of angular momentum can be expressed by setting the divergence of the expression under the integral sign in $M^{\mu\nu}$:

$$\frac{\partial}{\partial x^{\alpha}}(x^{\mu} T^{\nu\alpha} - x^{\nu} T^{\mu\alpha}) = 0 . \qquad (81)$$

Noting that $\partial x^{\mu}/\partial x^{\alpha} = \delta_{\alpha}{}^{\mu}$ and $\partial T^{\nu\alpha}/\partial x^{\alpha} = 0$, we find from Eq.(81)

$$\delta^{\mu}_{\alpha} T^{\nu\alpha} - \delta^{\nu}_{\alpha} T^{\mu\alpha} = T^{\nu\mu} - T^{\mu\nu} = 0$$

or

$$T^{\mu\nu} = T^{\nu\mu} , \qquad (82)$$

that is, conservation of angular momentum requires that the energy-momentum tensor must be symmetric.

Before we conclude this section, let us go back to Eq.(79). If we carry out the integration in Eq.(79) over the hyperplane x_0 = constant, then P^{μ} takes the form

$$P^{\mu} = \frac{1}{c}\int T^{\mu o} dV .$$

Here the integration extends over the whole three-dimensional space.

The space components of P^{μ} form the three-dimensional momentum vector of the system, and the time component is its energy multiplied by 1/c. Thus the vector with components

$$\frac{1}{c}T^{10}, \quad \frac{1}{c}T^{20}, \quad \frac{1}{c}T^{30}$$

may be called the "momentum density", and the quantity $W = T^{00}$ the "energy density".

Problems

(2-1) Observer 0 notes that two events are separated in space and time by 600 m and 8×10^{-7} s. How fast must an observer 0′ be moving relative to 0 in order that the events be simultaneous to 0′?

(2-2) A meter stick makes an angle of 30° with respect to x′-axis of 0′. What must be the value of v if the meter stick makes an angle of 45° with respect to the x-axis of 0?

(2-3) Find the speed of a particle that has a kinetic energy equal to exactly twice its rest mass energy.

(2-4) Find the law of transformation of the components of a symmetric four-tensor $T^{\mu\nu}$ under Lorentz transformations.

(2-5) A man on a station platform sees two trains approaching each other at the rate 7c/5, but the observer on one of the trains sees the other train approaching him with a velocity 35c/37. What are the velocities of the trains with respect to the station?

(2-6) The equation for a spherical pulse of light starting from the origin at t = t′ = 0 is

$$c^2dt^2 - x^2 - y^2 - z^2 = 0$$

Show from the Lorentz transformations that 0′ will observe this same pulse to be spherical, in accord with Einstein's postulate stating that the velocity of light is the same for all observers.

(2-7) Show that the Lorentz transformation matrix $a_{\mu\nu}$ satisfies the orthogonality conditions

$$\sum_{\lambda=0}^{3} a_{\lambda\mu} a_{\lambda\nu} = \delta_{\mu\nu}, \qquad \mu, \nu = 0,1,2,3 .$$

Chapter 3

Basic Ideas of General Relativity

3-1 *Inadequacy of special relativity*

Although special relativity rejects the ether concept of a privileged inertial frame of reference, it still strongly depends on the (Newtonian) concept of inertial frames. What is so special about these frames? This remained a mystery in special relativity, and it was reserved for general relativity to solve, or at least to elucidate, this problem. Einstein was led to general relativity primarily by his philosophic desire to abolish the role of absolute space from physics. Since special relativity begins by taking the inertial frames for granted, Einstein could never stop at special relativity, he had to go beyond it. In this respect, Einstein was deeply influenced by Ernst Mach (1836-1916). Mach's ideas on absolute space and inertia are roughly these: (a) space is simply the separation between bodies, and time is merely the succession of events; so neither space or time has an independent existence in their own right and relative motions are all that matter; (b) the property of inertia has nothing to do with absolute space, but arises from some kind of (unspecified) interaction between each individual body and all the other masses in the universe. If there were no other masses, an isolated body would have no inertia; this contrasted with Newton's view that the body would still have inertia because of the effect of absolute space. Einstein was very impressed by this whole complex of ideas expressed by Mach and coined the term **Mach's principle** to describe them. It should be noted that Mach's principle is rooted in classical kinematics, and it ignores "fields" as a possible content of space. Thus its very formulation within modern physics is questioned by many physicists. But Mach's principle provided a fruitful source of conjectures for Einstein to investigate quantitatively.

It is obvious that the concept of inertial frames of reference is incompatible with gravitational phenomena. An inertial frame is defined as one in which a free particle (i.e., no force acting on it) moves with a constant velocity. But gravity is long-range and cannot be screened.

Consequently, the only way to visualize an inertial frame is to imagine it far away from any matter. A concept like this is clearly of little use to someone doing experiments on earth or to astronomers whose observations relate to distant massive galaxies. Attempts to modify Newtonian gravity to make it compatible with special relativity have not been successful. According to Einstein, if gravitation is long-range and unscreened, it has something of a permanent character, and it must be intrinsic to the region in which it is located. Einstein identified this intrinsic property of a spacetime with its geometry. (We know that the geometry of spacetime is pseudo-Euclidean in special relativity.) In the presence of matter the spacetime geometry, according to Einstein, should be non-Euclidean. Therefore, he sought to relate the intrinsic parameters of a non-Euclidean spacetime geometry to the distribution of matter and energy. Thus, in general relativity the gravitational effects will not be described through an explicit external force, but rather through the non-Euclidean nature of the space-time geometry.

General relativity is unquestionably the modern theory of gravitation which has supplanted Newton's. Since both theories — Einsteinian special relativity and Newtonian gravitation — represent an approximation of the truth, we expect that general relativity should reduce to special relativity in cases where the gravitational effects are small, and should resemble Newtonian gravitation when special relativistic effects are negligible.

3-2 Einstein's principle of equivalence

Although we cannot introduce inertial frames of reference that are inertial in the strict Newtonian sense, in real cases we could cover space-time with a patchwork of local inertial frames.

The introduction of **local inertial frames** depends on the equivalence of inertial and gravitational mass. When gravity acts on a body, it acts on the gravitational mass m_g, and the result of the force is an acceleration of the inertial mass m_I. The fact that all bodies fall in a vacuum with the same acceleration indicates that within experimental accuracy, the ratio of inertial to gravitational mass is independent of the body. Newton realized this even when he formulated his laws of motion, and he was able to show that, if there was a difference between gravitational and inertial mass, it was not greater than one part in 1,000. If the fractional amount by which inertial and gravitational masses can differ is indicated by x, then Newton showed that $x < 1/1{,}000$. Bessels observed that the periods of pendulums made of different materials offered a sensitive test of the equivalence:

$$T = 2\pi \sqrt{\frac{m_I}{m_g} \frac{L}{g}} \; .$$

By measuring the periods (T) of pendulums of equal lengths (L) at the same place (the gravitational acceleration g is the same), he showed that

$$x < 1/(6 \times 10^4) \; .$$

Eötvos used a very ingenious method to set the limit more precisely. If a pendulum is set at a latitude λ as shown in Figure 1, the direction of

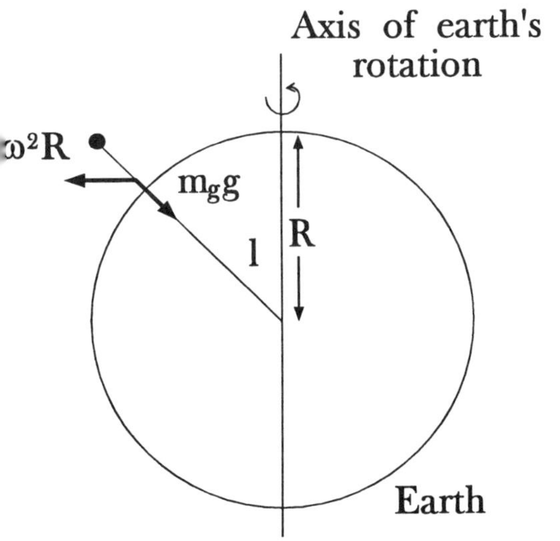

Figure 1: A pendulum at a latitude λ.

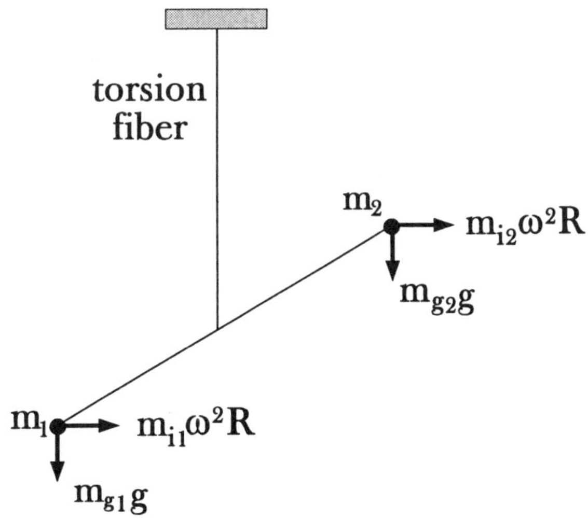

Figure 2: Eötvos's torsion pendulum.

the pendulum will not be toward the center of the earth but in the direction of the resultant of the forces $m_I\omega^2R$ and $m_g g$. If θ is the angle between the direction of the pendulum and the direction to the center of the earth, then θ is a simple function of m_I/m_g. Eötvos used a null method to compare the ratio m_I/m_g for different objects. His equipment consisted of a torsion pendulum (Figure 2). By orienting the axis m_1m_2 to the north, south, east, and west successively, and observing the twist in the torsion fiber with a mirror and microscope, any inequality of m_{I1}/m_{I2} for equal values of m_{g1}/m_{g2} may be determined. Eötvos established that x is less than (1/2) x 10^{-8}. Dicke redid the Eötvos experiment in 1964 using modern equipment and reported that $x < 1/10^{10}$. Braginski and Panov reported in 1971 that $x < 1/10^{12}$.

How can the equivalence of the inertial and gravitational mass of the same body make it possible to establish the **local inertial frames**? To see how, let us consider a region of space-time in which a constant gravitational field $\vec{\mathbf{g}}$ exists. One such region is the neighbourhood of any point on the surface of the earth. If gravity were the only force acting, all bodies in the region would fall with the same acceleration $\vec{\mathbf{a}} = \vec{\mathbf{g}}$. Thus, by transforming to a frame f' with acceleration $\vec{\mathbf{g}}$ we can eliminate the effects of gravitation; any object will appear unaccelerated unless a non-gravitational force $\vec{\mathbf{F}}_{ng}$ acts on it. It is easy to show this formally: consider a particle inside a stationary elevator cabin and it is acted upon by a gravitational force $m\vec{\mathbf{g}}$ and a non-gravitational $\vec{\mathbf{F}}_{ng}$; Newton's law gives

$$m\vec{\mathbf{a}} = m\vec{\mathbf{g}} + \vec{\mathbf{F}}_{ng} \quad .$$

If the elevator cabin is severed from its supporting cable and allowed to fall freely down a long shaft under the action of earth's gravity, in frame f' (the accelerated cabin) the acceleration of the particle relative to the cabin is $\vec{\mathbf{a}}' = \vec{\mathbf{a}} - \vec{\mathbf{g}}$, and so

$$m(\vec{\mathbf{a}}' + \vec{\mathbf{g}}) = m\vec{\mathbf{g}} + \vec{\mathbf{F}}_{ng}$$

or

$$m\vec{\mathbf{a}}' = \vec{\mathbf{F}}_{ng} \quad . \tag{1}$$

In Eq.(1), gravitational forces do not enter, i.e., gravity has been "transformed away" in the free-fall cabin (frame f'). Therefore, the equivalence of gravitational and inertial mass implies the following:

> in a small laboratory falling freely in a gravitational field, mechanical phenomena are the same as those observed in an inertial frame in the absence of a gravitational field.

In 1907 Einstein generalized this conclusion by replacing the word "mechanical phenomena" with "the laws of physics", and the resulting statement is known as the **principle of equivalence**.

Einstein's conclusion may at first seem difficult to accept. The principle seems to be derived from the Newtonian expression which we know is not rigorously valid. We do not derive it at all. We are only motivated towards it by the approximately valid Newtonian theory. Further, as the conclusion actually depends only on the equivalence of gravitational and inertial mass, it can be valid even when we reject the Newtonian mechanics. Einstein's equivalence principle, as above, is identical in content to the equivalence of inertial and gravitational mass. They are two different ways of formulating the same principle. Einstein's equivalence principle is sometimes called the "strong" equivalence principle, and the equivalence of inertial and gravitational mass is known as the "weak" equivalence principle.

Why must the laboratory be "small"? Because real gravitation fields are not constant: $\vec{\mathbf{g}}$ points towards the center of the gravitating matter, and so its direction varies from point to point round the gravitating matter, and its magnitude varies with height above the gravitating matter. Only over a sufficiently small region of the gravitational field can it be considered uniform.

These freely-falling frames covering the neighborhood of an event are called **local inertial frames**, and they are very important in relativity. Near an event there is an infinity of local inertial frames, each moving with constant velocity relative to the others. Special relativity applies rigorously within these frames, and the Lorentz transformation tells us how to transform the coordinates of an event from one of these frames to another.

It is obvious that local inertial frames are not inertial in the strict

Newtonian sense, because Eq.(1) only corresponds to Newton's second law if the gravitational force $m\vec{g}$ is ignored. They are more restricted and also more general than Newton's inertial frames: more restricted, because the inhomogeneity of real gravitational fields makes them only locally applicable, instead of infinite in extent; more general, because any laboratory in free fall (for instance **Skylab**) is a local inertial frame - it does not have to be unaccelerated relative to the galaxies, or to absolute space.

It is very important to bear in mind that we can only "transform away" gravitation in a limited region of spacetime by employing a laboratory in free-fall. This cannot be accomplished on a large scale — no laboratory exists which can cover all the space round a gravitating object and move in such a way as to eliminate its effects.

Can the effects of gravity also be created by choosing a suitable frame of reference? The answer is a definite yes! Consider an observer inside a closed cabin. When this cabin is fixed on earth surface, he will feel his normal weight, and will observe that all falling bodies accelerate towards the floor at the same rate (Figure 3a). Now if this observer were placed in an identical closed cabin out in space far from all massive bodies (so that no gravitational field exists), and if the cabin were fitted with a rocket motor capable of accelerating it smoothly at a rate exactly equal to the gravitational acceleration on earth, he would again find that all free objects accelerated towards the floor at the same rate, and he would also feel his normal weight (Figure 3b). There are no observations or experiments he could perform in his cabin which could tell him whether the effects were those of gravity or those of acceleration. Within a small closed cabin the effects of gravity and acceleration are indistinguishable.

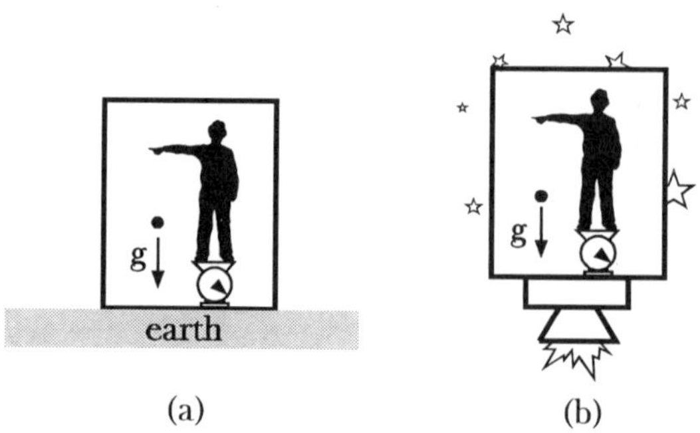

Figure 3: Equivalence of gravitation and acceleration.

Gravity in this context behaves like an **inertial force**, a fictitious force which arises due to the acceleration of the frame of reference from which observations are being made. The most familiar examples of inertial forces are centrifugal and Coriolis forces in a rotating system fixed on earth surface.

3-3 Immediate consequences of the principle of equivalence

The principle of equivalence leads very simply to two testable conclusions about the propagation of light. If the effects of gravitation and acceleration are indistinguishable, then rays of light should bend in a gravitational field. As well, light moving up through a gravitational field should be red-shifted. Let us look at these phenomena in turn.

A. The bending of a light beam

Consider an observer inside an enclosed spacecabin which is accelerating through a gravitation-free region of space. A ray of light enters the cabin from a window at P and it is parallel to the floor of the cabin as it enters the cabin. Where will the light hit on the opposite wall? The cabin will move upward while the light is travelling to the opposite wall so that the light will hit the wall at Q′, a little below Q. Thus, to an observer inside the cabin, the light ray curves downward as it travels through the cabin (Fig.4a). Now using the equivalence principle, we conclude that in a gravitaional field, light will not travel along a straight line, but its path will be curved (Fig. 4b).

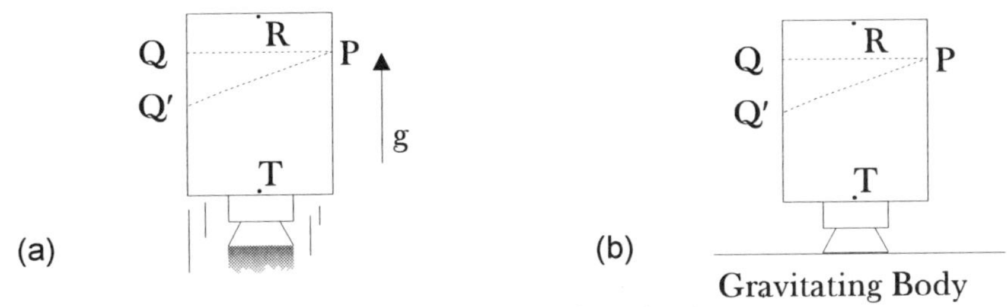

Figure 4: Equivalence principle predicts fall of light.

Near earth surface, where g is locally constant and parallel, this fall of light is far too small to be measurable; for instance, a horizontally-projected beam, after travelling 1 km, has fallen only about 1 Å. It is possible to detect the deflection of light falling past the Sun, but this involves the patching-together of many local inertial frames, and we put off the calculation for future work.

B. Gravitational shift of spectral lines
(Gravitational redshift)

Consider again the spacecabin, travelling in gravitation-free region of space with an acceleration $\vec{g}$. T and R be two fixed points on a straight line parallel to the direction of $\vec{g}$. A light wave of frequency ν is emitted by an atom at T. It will not have the same frequency when it reaches R. To analyze the situation we first note that the light ray will take a time $\Delta t = h/c$ to reach R (h is the distance between T and R) and during this time the point R gains an additional velocity $V = g\Delta t$. This apparent relative velocity with respect to T will cause a Doppler shift in frequency:

$$\nu' = \nu \frac{1 - \frac{V}{c}}{\sqrt{1 - \frac{V^2}{c^2}}} \simeq \nu\left(1 - \frac{gh}{c^2}\right) \tag{2}$$

Now using the equivalence principle, we can conclude that the same phenomenon musr be observed in a gravitational field. We note that gh is the difference in gravitational potential between source and receiver. If we call this difference $\Delta\phi$, we see that in passing up through a gravitational potential difference of $\Delta\phi$ the light has become redder and its wavelength is increased by the amount $\Delta\phi/c^2$. If light falls through a gravitational potential difference $\Delta\phi$, it gains energy and becomes bluer by the amount $\Delta\phi/c^2$, i.e. its wavelength is Doppler-shifted towards the blue by the amount $\Delta\phi/c^2$.

This remarkable prediction has been verified in a terrestrial laboratory in 1960 by Pound and Rebka. They allowed a 14.4 keV γ-ray, emitted by the radioactive decay of ^{57}Fe, to fall 22.6 m down an evacuated tower, and measured the change in its frequency. The predicted blue shift is z = -2.46 x 10^{-15}, and they measured z = (-2.57 ± 0.26) x 10^{-15}, thus directly verifying the equivalence principle. Such high precision was possible because of the Mossbauer effect; this is the emission of radiation from an atomic nucleus in a crystal, which gives a spectral line with a very precisely defined frequency.

The gravitational shift of spectral lines implies that in a gravitational field a clock (periodic phenomenon) runs slower than does the same clock in a gravitation-free region of space. We can regard a radiating atom as a clock, each "tick" being the emission of a wave crest. Since light from a clock in a gravitational field will be reddened when received by "clock at infinity", the latter clock will see the former clock ticking more slowly than itself if the clocks are of identical construction. The gravitational time dilation factor for a clock distant r from a mass M is

$$\Delta t(r) = (1 - \Delta\Phi/c^2)\Delta t = (1 - GM/rc^2)\Delta t \qquad (3)$$

where $\Delta t(r)$ and Δt are time intervals between events, measured in terms of ticks of the clocks at r and infinity.

3-4 The curved spacetime concept

Special relativity has familiarized us with space-time in terms of geometry, and the spacetime of special relativity is described by the Minkowskian metric

$$ds^2 = c^2 d\tau^2 = c^2 dt^2 - (dx^2 + dy^2 + dz^2)$$

where $d\tau$ is a measure of the proper time. Now since t(r) of Eq.(3) represents the time read by a clock at rest at r, this must be the proper time, that is for a clock at rest

$$ds^2 = c^2 d^2 = (1 - 2GM/rc^2)c^2 dt^2 \qquad (4)$$

apart from higher orders. Since the intervals of proper time are affected by gravitational field, we may say that the presence of a gravitational field influences the geometry of spacetime; and to the extent to which Eq.(4) is valid, we may write

$$ds^2 = (1 - 2GM/rc^2)c^2dt^2 - (dx^2 + dy^2 + dz^2) \tag{5}$$

and we see that the presence of the gravitational field modifies the structure of spacetime which is no longer Minkowskian. In fact, the proof of the existence of the gravitational red shift rules out the possibility of a theory of gravitation in Minkowski space. As soon as we assume that time is influenced by the gravitational field, we must also assume the possibility of its influencing the measure of length. For space and time coordinates must be treated on equal footing without any intrinsic preference of one over the other. (In fact, we expect the theory to reduce to Minkowski space in the limit of the gravitational field.) Thus, we may expect the general line element to be of the form

$$ds^2 = Ac^2dt^2 - (Bdx^2 + Cdy^2 + Ddz^2) \tag{6}$$

where A, B, C, and D depend on the gravitation and reduce to unity at large distance from the gravitating source. The coefficients A, B, C, and D are functions of the space-time variables x^o, x^1, x^2, x^3. But a gravitational field is equivalent to a certain noninertial frame of reference, and when we use a noninertial frame, the four-dimensional coordinate system is curvilinear. For example, if (x′, y′, z′, ct′) are the coordinates of a frame rotating uniformly with angular velocity ω along the z axis, the transformation to this frame is given by

$$x = x'\cos\omega t - y'\sin\omega t,\ y = x'\sin\omega t + y'\cos\omega t,\ z = z'$$

and the line element ds^2 becomes

$$ds^2 = [c^2 - \omega^2(x'^2 + y'^2)]dt^2 - dx'^2 - dy'^2 - dz'^2 + 2\omega y'\omega dx'dt - 2\omega x'dy'dt.$$

It is evident that this expression cannot be represented as a sum of squares of the coordinate differentials. Therefore, in a non-inertial frame of reference, the line-element is, in general, a quadratic form in the differentials of the coordinates x^0, x^1, x^2, and x^3 of the general type:

$$ds^2 = \sum_{\mu\nu} g_{\mu\nu} dx^\mu dx^\mu = g_{\mu\nu} dx^\mu dx^\nu \qquad (7)$$

where the $g_{\mu\nu}$ are functions of the space-time variables x^0, x^1, x^2, x^3, and are called the metric of the space-time manifold. The geometry of a space-time in which a metric form like Eq.(7) can be defined is called a Riemannian geometry.

As accelerated (noninertial) frames are equivalent to gravitational fields, gravitational effects are to be described by the metric $g_{\mu\nu}$. In this framework, gravitation is to be understood as a deviation of the metric of the space-time manifold from the flat Minkowski metric. Therefore, the metric $g_{\mu\nu}$ is not fixed arbitrarily on the whole space-time, but, as we will see, it depends on the local distribution of matter.

It is clear that $g_{\mu\nu}$ is symmetric in the indexes μ and ν ($g_{\mu\nu} = g_{\nu\mu}$). In the general case, there are ten different quantities $g_{\mu\nu}$ — four with equal indexes, and $4 \times 3/2 = 6$ with different indexes. In inertial frames, when we use Cartesian coordinates the quantities $g_{\mu\nu}$ are

$$g_{00} = 1,\ g_{11} = g_{22} = g_{33} = -1,\ \ g_{ik} = 0 \text{ for } i \neq k \qquad (8)$$

We call a four-dimensional system of coordinates with these values of $g_{\mu\nu}$ **galilean**. By an appropriate choice of coordinates, we can always bring the quantities $g_{\mu\nu}$ to galilean form at any individual point of the non-galilean space-time. We note that, after reduction to diagonal form at a given point, the matrix of the quantities $g_{\mu\nu}$ has one positive and three negative principal values. This set of signs is called the **signature** of the matrix. The determinant g, formed from the quantities $g_{\mu\nu}$, is always negative for a real space-time: $g < 0$.

We wish to stress a fundamental difference between "real" gravitational fields and noninertial frames, in spite of their local equivalence. A real gravitational field cannot be eliminated by any coordinate transformation. In other words, in the presence of a gravitational field, spacetime is such that the quantities $g_{\mu\nu}$ can not, by any coordinate transformation, be brought to their galilean values over all spacetime. Such a spacetime is said to be **curved**.

It is instructive to explain the idea of curvature. We have enough difficulty to imagine a curved three-dimensional space, let alone a curved four-dimensional spacetime, so we start with two-dimensional surfaces, with which we are all familiar. Obviously, a plane is flat. Consider a sphere or a cylinder; are these surfaces flat or curved? The sphere cannot be deformed to coincide with a plane without stretching or tearing, so a sphere is fundamentally curved. The curvature of the cylinder is less fundamental: it can be simply unrolled onto a plane without distortion.

In these three cases, we reached our conclusion by considering the three surfaces as embedded in three-dimensional position space. It was the great achievement of Gauss in the early nineteenth century to discover that the curvature and the whole geometry of a surface can be determined by employing the metric tensor $g_{\mu\nu}$. In a plane, the distance between two points separated by dx^1, dx^2 is

$$\begin{aligned} ds^2 &= g_{\mu\nu}dx^\mu dx^\nu \\ &= (dx^1)^2 + (dx^2)^2 \end{aligned} \tag{9}$$

where

$$g = \begin{pmatrix} 1 & 0 \\ 0 & 1 \end{pmatrix} . \tag{9a}$$

Pythagoras' theorem is satisfied, indicating that we are in a flat space. If we were given the metric distance in polar coordinates r, θ

$$ds^2 = dr^2 + r^2 d\theta^2$$

then the metric tensor is position-dependent

$$g = \begin{pmatrix} 1 & 0 \\ 0 & r^2 \end{pmatrix} .$$

How can we tell wether the surface is flat or not? By employing the following coordinate transformation

$$x^1 = r\cos\theta, \quad \text{and} \quad x^2 = r\sin\theta$$

we would get back the Cartesian distance and metric tensor formulas (9) and (9a). Now consider a cylinder of radius R. With cylindrical coordinates r, z, Φ in three-dimensional embedding space (Figure 5), the surface is defined by r = R = constant, and then

$$ds^2 = dz^2 + R^2 d\phi^2$$

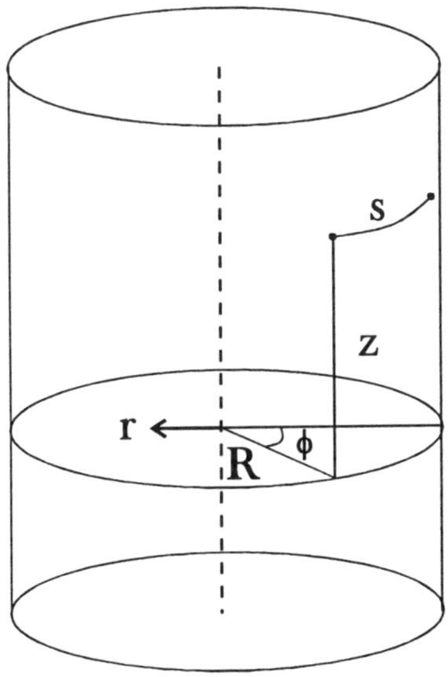

Figure 5: Line element on cylinder.

Now if we define coordinates x^1 and x^2:

$$x^1 = z,\ x^2 = R\phi, \quad \text{then} \quad ds^2 = (dx^1)^2 + (dx^2)$$

which is the same as the distance formula on a plane. Therefore the surface of a cylinder is intrinsically flat.

Next let us look at the surface of a sphere of radius R. We can use spherical coordinates θ and ϕ to describe positions on the surface of the sphere

$$x^1 = \theta,\ x^2 = \phi$$

Then we have

$$ds^2 = R^2 d\theta^2 + R^2 \sin^2\theta d\phi^2$$
$$= R^2(dx^1)^2 + R^2 \sin^2 x^1 (dx^2)^2$$

and the metric

$$g = \begin{pmatrix} R^2 & 0 \\ 0 & R^2 \sin^2 x^1 \end{pmatrix} .$$

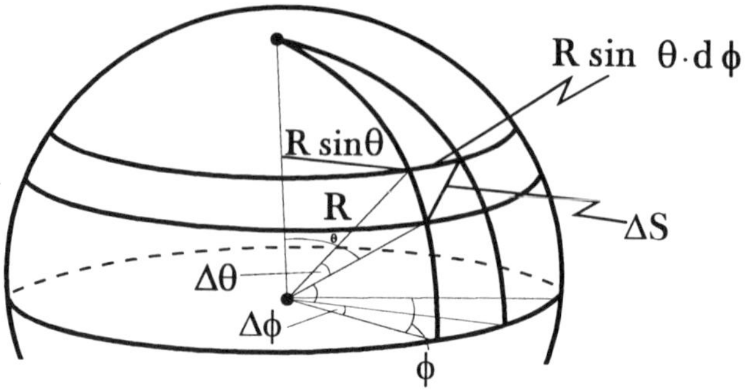

Figure 6: Line element on sphere.

Now we test for flatness by seeking new coordinates x'^1, x'^2, which are functions of x^1 and x^2 (i.e., of θ and ϕ); in terms of these new coordinates the distance must take the Cartesian form

$$(dx'^1)^2 + (dx'^2)^2 .$$

But however hard we try, we cannot find such a coordinate transformation, and it appears that the metrical properties of a spherical surface are intrinsically different from those of a plane. How can we know that there are no obscure transformations which will reduce a spherical surface to a plane? Can we know the curvature from the metric tensor alone? The answer was given by Gauss. He first showed that for the two-dimensional curved space the metric tensor can always be transformed to diagonal form (i.e., $g_{11} = g_{22} = 0$, and the metric is called orthogonal). He then showed that the curvature K of the surface is given by the formula

$$K = \frac{1}{2g_{11}g_{22}}\left\{ -\frac{\partial^2 g_{11}}{\partial (x^2)^2} - \frac{\partial^2 g_{22}}{\partial (x^1)^2} + \frac{1}{2g_{11}}\left[\frac{\partial g_{11}}{\partial x^1}\frac{\partial g_{22}}{\partial x^1} + \left(\frac{\partial g_{11}}{\partial x^2}\right)^2\right] + \frac{1}{2g_{22}}\left[\frac{\partial g_{11}}{\partial x^2}\frac{\partial g_{22}}{\partial x^2} + \left(\frac{\partial g_{22}}{\partial x^1}\right)^2\right]\right\} . \tag{10}$$

Obviously K is zero for a plane, either with Cartesian coordinates or polar coordinates; and $K = 1/R^2$ for the spherical surface of radius R. Unfortunately, in spaces of more than two dimensions it is not possible to specify curvature by only one function K. It turns out that a fourth-rank "curvature tensor" $R_{\mu\nu\lambda\beta}$ is necessary. It is defined in terms of derivatives of the metric tensor and the derivation is given in the following chapter on curvilinear coordinates and general tensors. For dealing with curved space we cannot introduce a rectilinear coordinate system. We have to use curvilinear coordinates.

3-5 *The principle of general covariance*

If we consider gravitational fields alone, the principle of equivalence denies us the possibility of distinguishing, by local measurements, between an inertial frame and a freely falling system in a gravitational field. There is then no a priori reason to give special status to inertial frames. For these and other reasons, Einstein was led to postulate that all frames of reference are equally good for the description of nature and that the laws of physics should have the same form in all. The equivalence of all frames of reference must be represented by the equivalence of all coordinate systems, since we have always represented frames of reference by coordinate systems. The requirement that, under a general coordinate transformation, the laws of physics must remain covariant (that is, form invariant) is called the **principle of general covariance**. It is the mathematical representation of the principle of equivalence. The treatment of generally covariant equations in a curved space is facilitated by using tensorial quantities for expreqsing the laws of physics. This can be seen from the following.

As will be demonstrated in the next chapter, if we have a tensor, say $T^{\mu\nu}$, defined in the coordinate system $S(x^0, x^1, x^2, x^3)$, the tensor will transform into

$$T'^{\mu\nu} = \frac{\partial x'^{\mu}}{\partial x^{\alpha}} \frac{\partial x'^{\nu}}{\partial x^{\beta}} T^{\alpha\beta}$$

under a coordinate transformation from S to S' (x'^0, x'^1, x'^2, x'^3). Since this is a completely linear form in the components of the tensor $T^{\mu\nu}$, the vanishing of all of its components in one coordinate system leads to the vanishing of all of its components in any other coordinate system. Therefore, let us consider, say in S, the tensor equation which represents a physics law,

$$A^{\mu\nu} = B^{\mu\nu} .$$

We now write

$$C^{\mu\nu} = A^{\mu\nu} - B^{\mu\nu}$$

and then, transforming this into coordinate system S', we have

$$C'^{\mu\nu} = \frac{\partial x'^{\mu}}{\partial x^{\alpha}} \frac{\partial x'^{\nu}}{\partial x^{\beta}} C^{\alpha\beta} .$$

As a consequence of the tensor equation, the vanishing of $C^{\mu\nu}$ in the unprimed coordinate system S leads to the vanishing of $C'^{\mu\nu}$ in the primed system

$$C'^{\mu\nu} = A'^{\mu\nu} - B'$$

or

$$A'^{\mu\nu} = B'^{\mu\nu}$$

that is, the physics law remains form invariant (covariant).

3-6 Distances and time intervals

In the theory of general relativity there exists no restriction of any kind regarding the choice of a permissible coordinate system. Thus, the question naturally arises that given a certain coordinate system $x^\mu(\mu = 0, 1, 2, 3)$, how do we relate the actual time and distance to these coordinates? We first find the relation of the proper time, which as before we shall denote by τ, to the coordinate x^0. In this respect let us consider two infinitesimally separated events. If these events take place at one and the same point in space, then the interval ds $(= g_{\mu\nu}\, dx^\mu\, dx^\nu)$ between the two events is just $cd\tau$, where $d\tau$ is the proper time interval between the two events. Stated differently, the proper time interval is the interval of time as measured by the clock of an observer who is at rest (at a point in space) in the gravitational field. Setting $dx^1 = dx^2 = dx^3 = 0$ in the general expression $ds = g^{\mu\nu}\, dx^\mu\, dx^\nu$, we get the connection between $d\tau$ (the element of proper time) and dx^0 (the coordinate differential):

$$d\tau = \frac{1}{c}\sqrt{g_{00}}dx^0 \tag{11}$$

or for the time between any two events occurring at the same point in space

$$\tau = \frac{1}{c}\int\sqrt{g_{00}}dx^0 \quad . \tag{12}$$

This relation determines the actual time interval (i.e., the proper time for the given point in space) for a change of the coordinate x^0.

We next determine the element $d\ell$ of spatial distance. In the special theory of relativity we can define $d\ell$ as the interval between two infinitesimally separated events occurring at one and the same time. However, we cannot define $d\ell$ in this way in the general theory of relativity, because in a gravitational field the proper time at different points in space has a different dependence on the coordinate x^0. To find $d\ell$, we can proceed as follows:

Consider two infinitesimally close points in space A and B, A has coordinates x^μ and B has coordinates $x^\mu + dx^\mu$. An observer at B sends a light signal to A and then receives it back over the same path in space (Figure 7). The time (as measured by the observer at B) required for this, when multiplied by c, is the distance between A and B. The interval ds between two events corresponding to the departure and arrival of a light signal from one to the other is equal to zero:

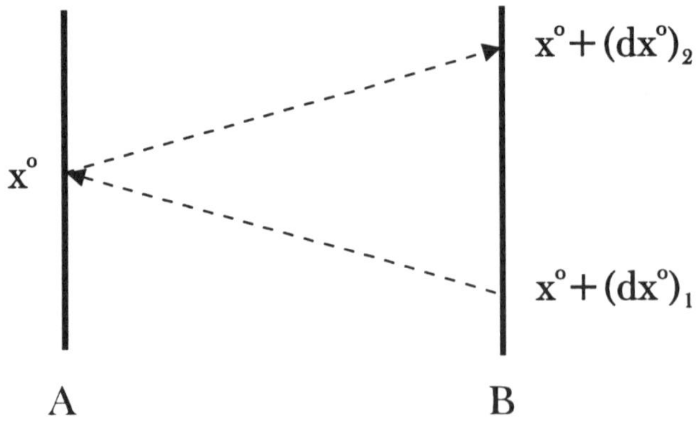

Figure 7: World lines of observers A and B, and the signals.

$$ds^2 = g_{\mu\nu}\, dx^\mu\, dx^\nu = 0 \tag{13}$$

or

$$g_{00}(dx^0)^2 + 2g_{0i}dx^i dx^0 + g_{ij}dx^i dx^j = 0 \quad i,j = 1,2,3 \tag{13a}$$

where we have separated the space and time coordinates. We find two roots for the coordinate time dx^0:

$$(dx^0)_1 = \frac{1}{g_{00}}\left[-g_{0i}dx^i - \sqrt{(g_{0i}g_{0j} - g_{ij}g_{00})dx^i dx^j}\right] \tag{14}$$

$$(dx^0)_2 = \frac{1}{g_{00}}\left[-g_{0i}dx^i + \sqrt{(g_{0i}g_{0j} - g_{ij}g_{00})dx^i dx^j}\right] . \tag{15}$$

If x^0 is the moment of arrival of the signal at A, the times when it left B and when it will return to B are, respectively, $x^0 + (dx^0)_1$ and $x^0 + (dx^0)_2$. In Figure 7 the solid lines are the world lines corresponding to the given coordinates x^μ and $x^\mu + dx^\mu$, and the dashed lines are the world lines of the signals. Therefore the total lapse of the coordinate time for the journey back and forth is equal to

$$(dx^0)_2 - (dx^0)_1 = \frac{2}{g_{00}}\sqrt{(g_{0i}g_{0j} - g_{ij}g_{00})dx^i dx^j} \tag{16}$$

and the corresponding interval of proper time is

$$\frac{1}{c}\sqrt{g_{00}} \times \left[(dx^0)_2 - (dx^0)_1\right] \tag{17}$$

We thus obtain, for the spatial distance $d\ell$ between two infinitesimally separated points, the expression

$$d\ell^2 = \left(-g_{ij} + \frac{g_{0i}g_{0j}}{g_{00}}\right)dx^i dx^j \ . \tag{18}$$

We can rewrite Eq.(18) in the form

$$d\ell^2 = \gamma_{ij}dx^i dx^j \tag{19}$$

where

$$\gamma_{ij} = \left(-g_{ij} + \frac{g_{0i}g_{0j}}{g_{00}}\right)dx^i dx^j \ . \tag{20}$$

The metric tensor $g_{\mu\nu}$ generally depends on x^0, so the space metric $d\ell^2$ also changes with time. For this reason, it is meaningless to integrate $d\ell$, as such an integral would depend on the world line chosen between the two given space points. Thus, generally speaking, in the general theory of relativity the concept of a definite distance between two bodies loses its meaning, remaining valid only for infinitesimal separations. Only when the metric tensor $g_{\mu\nu}$ does not depend on the time — and the distance can also then be defined over a finite domain — can the integral $d\ell$ along a space curve have a definite meaning.

Problems

(3-1) By direct transformation from Cartesian coordinates, calculate the metric tensor $g_{\mu\nu}$ in a flat three-dimensional space in terms of (a) spherical coordinates, and (b) cylindrical coordinates.

(3-2) Consider the red shift produced upon light escaping from the gravitational field of a celestial body. Show that the fractional change in wavelength of the light which escapes from the celestial body is

$$\frac{\Delta\lambda}{\lambda} = \frac{GM}{Rc^2} ,$$

where G = universal gravitational constant, R = radius of the celestial body, and M = mass.

(3-3) Using Gauss's curvature formula Eq.(10), show that K = 0 for a plane when plane polar coordinates are used, and that $K = 1/R^2$ for a sphere of radius a when spherical coordinates are used.

(3-4) Calculate the curvature K(r) of a surface whose metric distance formula is

$$\Delta s^2 = f(r)\Delta r^2 + r^2\Delta\phi^2 .$$

(3-5) Consider the gravitational red shift problem from the stand point of photons, and show that the fractional change of the frequency of the photon as it moves against the gravitation through a distance ℓ is

$$\frac{\Delta\nu}{\nu} = -\frac{g}{c^2} .$$

Chapter 4

Curvilinear Coordinates and General Tensors

4-1 Curvilinear coordinates

We devote this chapter to the development of four-dimensional geometry in arbitrary curvilinear coordinates. We shall deal with quantities which are located at a point in space. No such quantity will be disturbed by the curvature. Indeed, there may be a quantity of the same nature at all points of space; it then becomes a field quantity.

If we take a point quantity Q (or one of its components if it has several), we can differentiate with respect to any of the four coordinates. We write the result

$$\frac{\partial Q}{\partial x^{\mu}} = Q_{,\mu}$$

A subscript preceded by a comma will always denote a derivative in this way. We put the index downstairs in order that we may maintain a balancing of the indexes in the general equations. We can see this indexes balancing by noting that the change in Q, when we move from the point x^{μ} to a neighboring point $x^{\mu} + \delta x^{\mu}$, is

$$\delta Q = Q_{,\mu} \delta x^{\mu} \tag{1}$$

where summation over μ is understood. The repeated indexes appearing once in the lower and once in the upper position are automatically summed over (Einstein's summation convention).

Let us consider the transformation from one coordinate system x^0, x^1, x^2, x^3 to another x'^0, x'^1, x'^2, x'^3:

$$x^{\mu} = f^{\mu}(x'^0, x'^1, x'^2, x'^3) \tag{2}$$

where the f' are certain functions. When we transform the coordinates, their differentials transform according to the relation

$$dx^{\mu} = \frac{\partial x^{\mu}}{\partial x'^{\nu}} dx'^{\nu} . \tag{3}$$

Any set of four quantities A^{μ} ($\mu = 0,1,2,3$) which, under a coordinate change, transform like the coordinate differentials, is called a **contravariant vector**:

$$A^{\mu} = \frac{\partial x^{\mu}}{\partial x'^{\nu}} A'^{\nu} \tag{4}$$

If ϕ is some scalar, then under a coordinate change, the four quantities $\partial\phi/\partial x$ transform according to the formula

$$\frac{\partial \phi}{\partial x^{\mu}} = \frac{\partial \phi}{\partial x'^{\nu}} \frac{\partial x'^{\nu}}{\partial x^{\mu}} . \tag{5}$$

Any set of four quantities A_{μ} ($\mu = 0,1,2,3$) which, under a coordinate transformation, transform like the derivatives of a scalar is called a **covariant vector**:

$$A_{\mu} = \frac{\partial x'^{\nu}}{\partial x^{\mu}} A'_{\nu} \tag{6}$$

From the two contravariant vectors A^{μ} and B^{μ} we may form the 16 quantities $A^{\mu}B^{\mu}$ ($\mu, \nu = 0,1,2,3$). These 16 quantities form the components of a contravariant tensor of the second rank: Any aggregate of 16 quantities $T^{\mu\nu}$ which, under a coordinate transformation, transform like the products of two contravariant vectors

$$T^{\mu\nu} = \frac{\partial x^{\mu}}{\partial x'^{\alpha}} \frac{\partial x^{\nu}}{\partial x'^{\beta}} T'^{\alpha\beta} \tag{7}$$

is a contravariant tensor of rank two. We may also form a **covariant tensor** of rank two from two covariant vectors, which transform according to the formula

$$T_{\mu\nu} = \frac{\partial x'^{\alpha}}{\partial x^{\mu}} \frac{\partial x'^{\beta}}{\partial x^{\nu}} T'_{\alpha\beta} \tag{8}$$

Similarly, we can form a mixed tensor $T_{\nu\mu}$ of order two which transforms as follows

$$T^{\mu}{}_{\nu} = \frac{\partial x^{\mu}}{\partial x'^{\alpha}}\frac{\partial x'^{\beta}}{\partial x^{\nu}}T'^{\alpha}{}_{\beta} \ . \tag{9}$$

We may continue this process and multiply more than two vectors together, taking care that their indexes are all different. In this way we can construct tensors of higher rank. The total number of free indexes of a tensor is called its rank (or order).

We may set a subscript equal to a superscript and sum over all values of this index, which results in a tensor having two fewer free indexes than the original one. This process is called **contraction**. For example, if we start with a fourth-order tensor $T^{\mu}{}_{\nu\rho}{}^{\sigma}$, one way of contracting it is to put $\sigma = \rho$, which gives the second rank tensor $T^{\mu}{}_{\nu\rho}{}^{\rho}$, having only 16 components, arising from the four values of μ and ν. We could contract again to get the scalar $T^{\mu}{}_{\mu\rho}{}^{\rho}$, with just one component.

It is easy to show that the inner product of contravariant and covariant vectors, $A_{\mu}B^{\mu}$, is an invariant, that is, independent of the coordinate system

$$A'_{\mu}B'^{\mu} = \frac{\partial x^{\alpha}}{\partial x'^{\mu}}\frac{\partial x'^{\mu}}{\partial x^{\beta}}A_{\alpha}B^{\beta} = \frac{\partial x^{\alpha}}{\partial x^{\beta}}A_{\alpha}B^{\beta} = \delta^{\alpha}{}_{\beta}A_{\alpha}B^{\beta} = A_{\alpha}B^{\alpha} \ .$$

The square of the line element in curvilinear coordinates is a quadratic form in the differentials dx^{μ}: $ds^2 = g_{\mu\nu}dx^{\mu}dx^{\nu}$. Since the (contracted) product of $g_{\mu\nu}$ and the contravariant tensor $dx^{\mu}dx^{\nu}$ is a scalar, the $g_{\mu\nu}$ forms a covariant tensor:

$$ds^2 = g'_{\alpha\beta}dx'^{\alpha}dx'^{\beta} = g_{\alpha\beta}dx^{\alpha}dx^{\beta} \ .$$

Now $dx'^{\alpha} = (\partial x'^{\alpha}/\partial x^{\mu})dx^{\mu}$, so that

$$g'_{\alpha\beta}\frac{\partial x'^{\alpha}}{\partial x^{\mu}}\frac{\partial x'^{\beta}}{\partial x^{\nu}}dx^{\mu}dx^{\nu} = g_{\mu\nu}dx^{\mu}dx^{\nu}$$

or

$$\left(g'_{\alpha\beta}\frac{\partial x'^{\alpha}}{\partial x^{\mu}}\frac{\partial x'^{\beta}}{\partial x^{\nu}} - g_{\mu\nu}\right)dx^{\mu}dx^{\nu} = 0 \ .$$

The above equation is identically zero for arbitrary dx^{μ}, so we have

$$g_{\mu\nu} = \frac{\partial x'^{\alpha}}{\partial x^{\mu}}\frac{\partial x'^{\beta}}{\partial x^{\nu}}g'_{\alpha\beta} \tag{10}$$

that is, $g_{\mu\nu}$ is a covariant tensor of rank two. It is called the **metric tensor** or the **fundamental tensor**.

So far covariant and contravariant vectors have no direct connection with each other except that their inner product is an invariant. A space in which covariant and contravariant vectors exist separately is called **affine**. Physical quantities are independent of the particular choice of the mode of description, that is, independent of the possible choices of contravariance or covariance. In metric space, contravariant and covariant vectors can be converted into each other with the help of the fundamental tensor $g_{\mu\nu}$. For example, we can get the covariant vector A_μ from the contravariant vector A^ν

$$A_\mu = g_{\mu\nu}A^\nu \; . \tag{11}$$

Since the determinant $|g|$ does not vanish, these equations can be solved for A^ν in terms of the A_μ. Let the result be

$$A^\nu = g^{\mu\nu}A_\mu \; . \tag{12}$$

By combining the two transformations (11) and (12), we have

$$A_\mu = g_{\mu\nu}g^{\nu\alpha}A_\alpha \; .$$

Since the equation must hold for any four quantities A_μ, we can infer

$$g_{\mu\nu}g^{\nu\alpha} = \delta^\alpha_\mu \tag{13}$$

In other words, $g^{\mu\nu}$ is the inverse of $g_{\mu\nu}$ and vice versa, that is

$$g^{\mu\nu} = M^{\mu\nu}/|g| \tag{14}$$

where $|g|$ is the determinant of $g_{\mu\nu}$ and $M^{\mu\nu}$ is the minor of the element $g_{\mu\nu}$.

Eq.(11) may be used to lower any upper index occurring in a tensor. Similarly, Eq.(12) can be used to raise any downstairs index. It is necessary to remember the position from which the index was lowered or raised, because when we bring the index back to its original site, we do not want to interchange the order of indexes, in general $T^{\mu\nu} \neq T^{\nu\mu}$.

Two tensors $A_{\mu\nu}$ and $B^{\mu\nu}$ are said to be reciprocal to each other if

$$A_{\mu\nu}B^{\nu\alpha} = \delta^\alpha_\mu \; . \tag{15}$$

A tensor is called symmetric with respect to two contravariant or two covariant indexes if its components remain unaltered on interchange of the indexes. Thus if $A^{\mu\nu\alpha}_{\beta\gamma} = A^{\nu\mu\alpha}_{\beta\gamma}$ the tensor is symmetric in μ and ν. If a tensor is symmetric with respect to any two contravariant and any two covariant indices, it is called symmetric.

A tensor is called skew-symmetric with respect to two contravariant or two covariant indexes if its components change sign upon interchange of the indexes. Thus if $A^{\mu\nu\alpha}_{\beta\gamma} = -A^{\nu\mu\alpha}_{\beta\gamma}$ the tensor is skew- symmetric in μ and ν. If a tensor is skew-symmetric with respect to any two contravariant and any two covariant indexes, it is called skew-symmetric.

If a tensor is symmetric (or skew-symmetric) with respect to two indexes in one coordinate system, it remains symmetric (skew-symmetric) with respect to those two indexes in any other coordinate system. It is easy to prove this. For example, if $B^{\alpha\beta}$ is symmetric, $B^{\alpha\beta} = B^{\beta\alpha}$, then

$$B'^{\alpha\beta} = \frac{\partial x'^{\alpha}}{\partial x^{\gamma}}\frac{\partial x'^{\beta}}{\partial x^{\delta}} B^{\gamma\delta} = \frac{\partial x'^{\alpha}}{\partial x^{\gamma}}\frac{\partial x'^{\beta}}{\partial x^{\delta}} B^{\delta\gamma} = B^{\beta\alpha} \tag{16}$$

and $B^{\alpha\beta}$ remains symmetric in the x′ coordinate system.

Every tensor can be expressed as the sum of two tensors, one of which is symmetric and the other skew-symmetric in a pair of covariant or contravariant indices. Consider, for example, the tensor $B^{\alpha\beta}$. We can write it as

$$B^{\alpha\beta} = \frac{1}{2}(B^{\alpha\beta} + B^{\beta\alpha}) + \frac{1}{2}(B^{\alpha\beta} - B^{\beta\alpha}) \, . \tag{17}$$

It is easy to see that the first term on the right-hand side is symmetric, and the second term is skew-symmetric. By similar reasoning the result is seen to be true for any tensor.

It is obvious that the sum or difference of two or more tensors of the same rank and type (i.e., same number of contravariant indices and same number of covariant indices) is also a tensor of the same rank and type. Thus if $A^{\mu\nu}_{\lambda}$ and $B^{\mu\nu}_{\lambda}$ are tensors, then $C^{\mu\nu}_{\lambda} = A^{\mu\nu}_{\lambda} + B^{\mu\nu}_{\lambda}$ and $D^{\mu\nu}_{\lambda} = A^{\mu\nu}_{\lambda} - B^{\mu\nu}_{\lambda}$ are also tensors.

A given quantity $N^{\mu\dots}_{\nu\rho\dots}$ with various up and down indexes may or may not be a tensor. We can test whether it is a tensor or not by using the **quotient law,** which can be stated as follows.

> *Suppose it is not known whether a quantity X is a tensor or not. If an inner product of X with an arbitrary tensor is a tensor, then X is also a tensor.*

As an example, let $X = P_{\lambda\mu\nu}$, A^{λ} is an arbitrary contravariant vector, and $A^{\lambda}P_{\lambda\mu\nu}$ is a tensor, say $Q_{\mu\nu}$: $A^{\lambda}P_{\lambda\mu\nu} = Q_{\mu\nu}$, then

$$A^{\lambda}P_{\lambda\mu\nu} = \frac{\partial x'^{\alpha}}{\partial x^{\mu}}\frac{\partial x'^{\beta}}{\partial x^{\nu}}A'^{\gamma}P'_{\gamma\alpha\beta} .$$

But

$$A'^{\gamma} = \frac{\partial x'^{\gamma}}{\partial x^{\lambda}}A^{\lambda} .$$

So

$$A^{\lambda}P_{\lambda\mu\nu} = \frac{\partial x'^{\alpha}}{\partial x^{\mu}}\frac{\partial x'^{\beta}}{\partial x^{\nu}}\frac{\partial x'^{\gamma}}{\partial x^{\lambda}}A'^{\lambda}P'_{\gamma\alpha\beta} .$$

This equation must hold for all values of A^{λ}, so

$$P_{\lambda\mu\nu} = \frac{\partial x'^{\alpha}}{\partial x^{\mu}}\frac{\partial x'^{\beta}}{\partial x^{\nu}}\frac{\partial x'^{\gamma}}{\partial x^{\lambda}}P'_{\gamma\alpha\beta} \tag{18}$$

showing that $P_{\lambda\mu\nu}$ is a tensor (contravariant tensor of rank 3).

For a nontensor $N^{\mu}_{\nu\eta}$ we can raise and lower indexes by the same rules as for a tensor. Thus for example

$$g^{\alpha\nu}N^{\mu}{}_{\nu\eta} = N^{\mu\alpha}{}_{\eta} . \tag{19}$$

4-2 Parallel displacement and covariant differentiation

We have seen that a covariant vector is transformed according to the formula

$$A_i = \frac{\partial x'^k}{\partial x^i}A'_k \tag{20}$$

where the coefficients are functions of the coordinates. So vectors at different points transform differently. Because of this fact, dA_i is not a vector, since it is the difference of vectors located at two (infinitesimally separated) points of space-time. We can easily verify this directly from Eq.(20)

$$dA_i = \frac{\partial x'^k}{\partial x^i} dA'_k + A'_k \frac{\partial^2 x'^k}{\partial x^i \partial x^j} dx^j$$

which shows that dA_i does not transform at all like a vector. The same also applies to the differential of a contravariant vector.

When using curvilinear coordinates, a differential can be obtained only when the two vectors to be subtracted from each other are located at the same point in space-time. In order to do so, we must parallel displace one of the vectors to the point where the other vector is located, after which we determine the difference of two vectors which now refer to one and the same point in space-time.

The concept of parallel displacement of a vector is very clear in Cartesian coordinates: displace a vector parallel to itself so that both its length and orientation are unchanged. We can extend the idea of parallel displacement of a vector to curved spaces in a consistent way. This requires us to assume that there always exist Galilean coordinates in the immediate vicinity of a point in space-time; in such a coordinate system the idea of an infinitesimal parallel displacement of a vector works, in other words, a vector can be transported parallel to itself without changing its length and orientation. This is illustrated, in Figure 1, with the example of a curved two-dimensional surface in a three-dimensional Euclidean space. As a vector and none of its components change under parallel displacement, the inner (scalar) product of two vectors $\mathbf{A}^\mu$ and $\mathbf{B}^\mu$, $A_\mu B^\mu$, does not change under parallel displacement. For arbitrary coordinates we **define** the operation of infinitesimal parallel displacement of a vector A^μ from a point P to a neighboring point Q to be one which leaves the inner product with an arbitrary vector B_μ invariant.

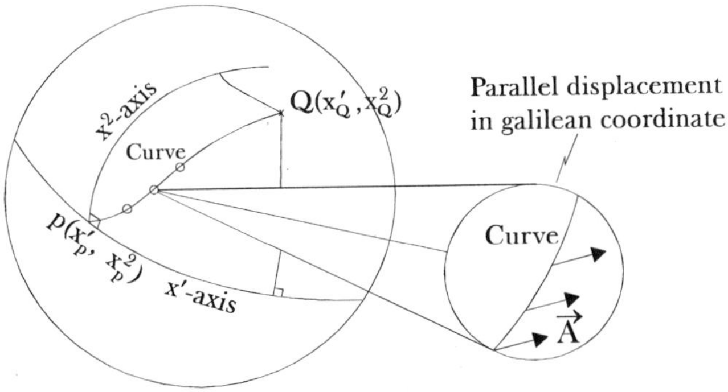

Figure 1: Parallel displacement in curvilinear coordinates.

Parallel displacement is independent of the path taken on a Euclidean plane (a flat surface), as shown in Figure 2a. On a curved surface, however, we will obtain a different final result depending on the path taken (Figure 2b).

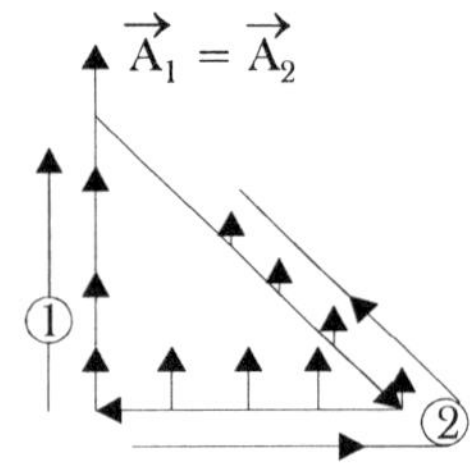

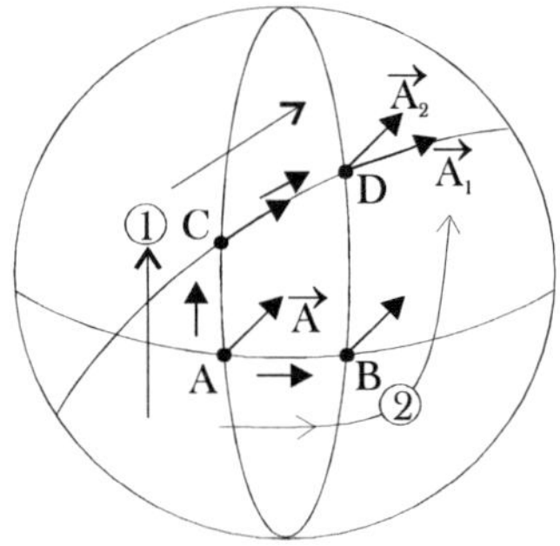

Figure 2: Parallel transport around a closed curve.

We can transfer a vector continuously along a path by the process of parallel displacement. In curvilinear coordinates, the components of a vector would be expected to change under a parallel displacement, unlike the case of a Cartesian coordinate. Therefore, if A^μ are the components of a contravariant vector at the point $P(x^\mu)$, and $A^\mu + dA^\mu$ the components at a neighboring point $Q(x^\mu + dx^\mu)$, where

$$A^\mu + dA^\mu = A^\mu + \frac{\partial A^\mu}{\partial x^\sigma} dx^\sigma , \tag{21}$$

an infinitesimal parallel displacement of A^μ from P to Q would produce a variation of its components, δA^μ. δA^μ should be a linear function of the coordinate differentials and the components A^μ. We write it in the form

$$\delta A^\nu = -\Gamma^\nu_{\alpha\beta} A^\alpha dx^\beta \tag{22}$$

where the $\Gamma^\nu_{\alpha\beta}$ are certain functions of the coordinates and are called Christoffel symbols of the second kind. Their form depends on the coordinate system. It will be proved in Section 4-4 that in a Galilean coordinate system $\Gamma^\nu_{\alpha\beta} = 0$. From this it is already clear that the quantities $\Gamma^\nu_{\alpha\beta}$ do not form a tensor, since a tensor which is equal to zero in one coordinate system is equal to zero in every other one. In a curved space it is impossible to make all the $\Gamma^{\nu}_{\alpha\beta}$ vanish over all of space.

The vector resulting from parallel displacement from point P to point Q is $A^{\mu} + \delta A^{\mu}$. Subtracting these two quantities gives

$$DA^{\mu} = dA^{\mu} - \delta A^{\mu} = \left(\frac{\partial A^{\mu}}{\partial x^{\sigma}} + \Gamma^{\mu}{}_{\sigma\alpha} A^{\alpha} \right) dx^{\sigma} . \tag{23}$$

We would expect the difference $dA^{\mu} - \delta A^{\mu}$ to be a vector since it is the difference of two vectors at the <u>same point</u>; the quantity

$$\frac{\partial A^{\mu}}{\partial x^{\sigma}} + \Gamma^{\mu}{}_{\sigma\alpha} A^{\alpha}$$

then, is a mixed tensor called the **contravariant derivative** of A^{μ} and written

$$A^{\mu}{}_{;\sigma} = \frac{\partial A^{\mu}}{\partial x^{\sigma}} + \Gamma^{\mu}{}_{\sigma\alpha} A^{\alpha} . \tag{24}$$

From $\delta(A_{\mu}A^{\mu}) = 0$ it follows, using Eq.(22), that

$$\delta A_{\mu} = \Gamma^{\alpha}{}_{\mu\beta} A_{\alpha} dx^{\beta} .$$

From this and a similar procedure that leads to Eqs.(23) and (24) we obtain the **covariant derivative** of A_{μ}:

$$A_{\mu;\sigma} = \frac{\partial A_{\mu}}{\partial x^{\sigma}} - \Gamma^{\alpha}{}_{\mu\sigma} A_{\alpha} . \tag{25}$$

The tensor character of Eqs.(24) and (25) can be established formally by showing that they obey the required transformation laws. This will require us first to establish the transformation laws for the $\Gamma^{\alpha}_{\mu\delta}$; it is not difficult to do this, but it is very tedious. Because of limited space-time in this book we shall leave this out.

The **contravariant derivative** is formed by raising the index which denotes differentiation,

$$A^{\mu;\sigma} = g^{\sigma\alpha} A^{\mu}{}_{;\alpha} . \tag{26}$$

In Galilean coordinates, $\Gamma^{\alpha}_{\mu\beta} = 0$, and covariant differentiation reduces to ordinary differentiation.

We may also obtain the covariant derivative of a tensor by determining the change in the tensor under an infinitesimal parallel displacement. For example, let us consider any arbitrary tensor $T^{\mu\nu}$, expressible as a product of two contravariant vectors $A^{\mu}B^{\nu}$. Under infinitesimal parallel displacement

$$\delta(A^{\mu}B^{\nu}) = A^{\mu}\delta B^{\nu} + B^{\nu}\delta A^{\mu} = -A^{\mu}\Gamma^{\nu}{}_{\alpha\beta}B^{\alpha}dx^{\beta} - B^{\nu}\Gamma^{\mu}{}_{\beta\alpha}A^{\alpha}dx^{\beta} .$$

By virtue of the linearity of this transformation we also have

$$\delta T^{\mu\nu} = -\left(T^{\mu\beta}\Gamma^{\nu}{}_{\beta\alpha} + A^{\beta\nu}\Gamma^{\mu}{}_{\beta\alpha}\right)dx^{\alpha} .$$

Substituting this in

$$DA^{\mu\nu} = dA^{\mu\nu} - \delta A^{\mu\nu} = A^{\mu\nu}{}_{;\alpha}dx^{\alpha}$$

we get the covariant derivative of the tensor $T^{\mu\nu}$ in the form

$$T^{\mu\nu}{}_{;\alpha} = \frac{\partial T^{\mu\nu}}{\partial x^{\alpha}} + \Gamma^{\mu}{}_{\beta\alpha}T^{\beta\nu} + \Gamma^{\nu}{}_{\beta\alpha}T^{\mu\beta} . \qquad (27)$$

In similar fashion we obtain the covariant derivative of the mixed tensor $T^{\mu}{}_{\nu}$ and the covariant tensor $T_{\nu\mu}$ in the form

$$T^{\mu}{}_{\nu;\alpha} = \frac{\partial T^{\mu}{}_{\nu}}{\partial x^{\alpha}} - \Gamma^{\beta}{}_{\nu\alpha}T^{\mu}{}_{\beta} + \Gamma^{\mu}{}_{\beta\alpha}T^{\beta}{}_{\nu} , \qquad (28)$$

$$T_{\mu\nu;\alpha} = \frac{\partial T_{\mu\nu}}{\partial x^{\alpha}} - \Gamma^{\beta}{}_{\mu\alpha}T_{\beta\nu} - \Gamma^{\beta}{}_{\nu\alpha}T_{\mu\beta} . \qquad (29)$$

One can similarly determine the covariant derivative of a tensor of arbitrary rank. In doing this one finds the following rule of covariant differentiation:

> *To obtain the covariant derivative of the tensor* $T^{\cdots\cdots}_{\cdots\cdots}$ *with respect to* x^{μ}, *you add to the ordinary derivative* $\partial T\ldots/\partial x^{\mu}$ *for each covariant index* $\nu\left(T^{\cdots\cdots}_{..\nu..}\right)$ *a term* $-\Gamma^{\alpha}_{\mu\nu}T^{\cdots\cdots}_{..\alpha..}$, *and for each contravariant index* $\nu\left(T^{\ldots\nu.}_{\cdots\cdots}\right)$ *a term* $+\Gamma^{\alpha}_{\nu\mu}T^{\ldots\alpha.}_{\cdots\cdots}$.

The covariant derivative of the metric tensor $g_{\mu\nu}$ is zero. To show this we note that the relation

$$DA_\mu = g_{\mu\nu} DA^\nu$$

is valid for the vector DA_μ as for any vector. On the other hand, we have $A_\mu = g_{\mu\nu}A^\nu$, so that

$$DA_\mu = D(g_{\mu\nu}A^\nu) = g_{\mu\nu}DA^\nu + A^\nu Dg_{\mu\nu} .$$

Comparing with $DA_\mu = g_{\mu\nu}DA^\nu$, we have $A^\nu Dg_{\mu\nu} = 0$. But the vector A^ν is arbitrary, so

$$Dg_{\mu\nu} = 0 .$$

Therefore the covariant derivative

$$g_{\mu\nu;\alpha} = 0 . \tag{30}$$

Thus $g_{\mu\nu}$ may be considered as a constant during covariant differentiation.

The covariant derivative of a product can be found by the same rule as for ordinary differentiation of products. In doing this we must consider the covariant derivative of a scalar ϕ as an ordinary derivative, that is, as the covariant vector $\phi_k = \partial\phi/\partial x^k$, in accordance with the fact that for a scalar $\delta\phi = 0$, and hence $D\phi = d\phi$. For example

$$(A_\mu B\nu)_{;\alpha} = A_{\mu;\alpha}B_\nu + A_\mu B_{\nu;\alpha} . \tag{31}$$

4-3 Symmetry properties of the Christoffel symbols

We now show that $\Gamma^\nu_{\alpha\beta}$ is symmetric in the subscripts. Let δA^ν be a coordinate differential dx^ν, then Eq.(22) becomes

$$\delta(dx^\nu) = -\Gamma^\nu{}_{\alpha\beta}dx^\alpha dx^\beta . \tag{32}$$

Next, we return to the local Cartesian coordinate system by the transformations

$$\left.\begin{aligned} x^\alpha &= f'(x'^1, x'^2, \ldots.) \\ x'^\alpha &= \varphi^\alpha(x^1, x^2, \ldots\ldots) \end{aligned}\right\} \tag{33}$$

where the primed coordinates are local Cartesian coordinates. From Eq.(33) we obtain

$$dx^{\alpha} = \frac{\partial f^{\alpha}}{\partial x'^{\beta}} dx'^{\beta} \tag{34}$$

Under a parallel displacement, $\delta(dx'^{\beta}) = 0$ by definition, so that from Eq.(34) we have

$$\delta(dx^{\nu}) = \frac{\partial^2 f^{\nu}}{\partial x'^{\delta} \partial x'^{\gamma}} dx'^{\delta} dx'^{\gamma} = \frac{\partial^2 f^{\nu}}{\partial x'^{\delta} \partial x'^{\gamma}} \frac{\partial \varphi^{\delta}}{\partial x^{\alpha}} \frac{\partial \varphi^{\gamma}}{\partial x^{\beta}} dx^{\alpha} dx^{\beta} .$$

Comparing this with Eq.(32) we obtain

$$\Gamma^{\nu}{}_{\alpha\beta} = - \frac{\partial^2 f^{\nu}}{\partial x'^{\delta} \partial x'^{\gamma}} \frac{\partial \varphi^{\delta}}{\partial x^{\alpha}} \frac{\partial \varphi^{\gamma}}{\partial x^{\beta}} . \tag{35}$$

The right-hand side of Eq.(35) is clearly symmetric in the indexes α and β, so that $\Gamma^{\nu}_{\alpha\beta}$ must be symmetric in α and β also.

4-4 Christoffel symbols and the metric tensor

It is very useful to express the Γ's in terms of the metric tensors. Let A^{μ} be any contravariant vector, $A_{\mu} = g_{\mu\nu} A^{\nu}$ a covariant vector. From the definition of parallel displacement $\delta(A_{\mu} A^{\mu}) = 0$, we have

$$\delta(A_{\mu} A^{\mu}) = g_{\mu\nu}(x^{\mu} + dx^{\mu})[A^{\nu} + \delta A^{\nu}][A^{\mu} + \delta A^{\mu}] - g_{\mu\nu}(x^{\mu}) A^{\nu} A^{\mu} = 0.$$

Carrying out these operations gives

$$\frac{\partial g_{\mu\nu}}{\partial x^{\alpha}} A^{\mu} A^{\nu} dx^{\alpha} + g_{\mu\nu} A^{\mu} \delta A^{\nu} + g_{\mu\nu} A^{\nu} \delta A^{\mu} = 0$$

Making use of Eq.(22) to eliminate δA^{μ} and δA^{μ} gives

$$\frac{\partial g_{\mu\nu}}{\partial x^{\alpha}} - g_{\mu\beta} \Gamma^{\beta}{}_{\nu\alpha} - g_{\nu\beta} \Gamma^{\beta}{}_{\mu\alpha} = 0 . \tag{36}$$

As shown above, $\Gamma^{\beta}_{\nu\alpha}$ is symmetric in the lower indexes ν and α. This symmetry allows permutation of ν and α to obtain

$$\frac{\partial g_{\mu\alpha}}{\partial x^{\nu}} - g_{\mu\beta}\Gamma^{\beta}{}_{\nu\alpha} - g_{\alpha\beta}\Gamma^{\beta}{}_{\mu\nu} = 0 \ . \tag{37}$$

Similarly we write

$$\frac{\partial g_{\nu\alpha}}{\partial x^{\mu}} - g_{\nu\beta}\Gamma^{\beta}{}_{\mu\alpha} - g_{\alpha\beta}\Gamma^{\beta}{}_{\mu\nu} = 0 \ . \tag{38}$$

Solving Eqs.(36), (37), (38) for $\Gamma^{\gamma}_{\mu\alpha}$, with the help of $g_{\mu\nu}g^{\nu\alpha} = \delta^{\alpha}_{\mu}$:

$$\Gamma^{\gamma}_{\mu\alpha} = \frac{1}{2}g^{\gamma\nu}\left[\frac{\partial g_{\nu\mu}}{\partial x^{\alpha}} + \frac{\partial g_{\nu\alpha}}{\partial x^{\mu}} - \frac{\partial g_{\mu\alpha}}{\partial x^{\nu}}\right] . \tag{39}$$

The Christoffel symbol of the first kind is

$$\Gamma_{\nu,\mu\alpha} = \frac{1}{2}\left[\frac{\partial g_{\nu\mu}}{\partial x^{\alpha}} + \frac{\partial g_{\nu\mu}}{\partial x^{\mu}} - \frac{\partial g_{\mu\alpha}}{\partial x^{\nu}}\right] \tag{40}$$

and it is often written as $[\mu\alpha, \nu]$. Clearly, $\Gamma_{\nu,\mu,\alpha} = \Gamma_{\nu,\alpha,\mu}$. Christoffel symbols are also known as affine connections. Since the metric tensors are all constant in Galilean coordinates, the Christoffel symbols all vanish.

The equation $g_{\mu\nu;\sigma} = 0$ can be used to offer an alternative derivation for the expressions for the Christoffel symbols in terms of the metric tensor, Eqs.(39) and (40). To this end we write in accordance with the general definition (29):

$$g_{\mu\nu;\alpha} = \frac{\partial g_{\mu\nu}}{\partial x^{\alpha}} - g_{\beta\nu}\Gamma^{\beta}{}_{\mu\alpha} - g_{\mu\beta}\Gamma^{\beta}{}_{\nu\alpha} = \frac{\partial g_{\mu\nu}}{\partial x^{\alpha}} - \Gamma_{\nu,\mu\alpha} - \Gamma_{\mu,\nu\alpha} = 0 \ .$$

From this we have, permuting the indexes μ, ν, α:

$$\frac{\partial g_{\mu\nu}}{\partial x^{\alpha}} = \Gamma_{\nu,\mu\alpha} + \Gamma_{\mu,\nu\alpha} \ .$$

$$\frac{\partial g_{\alpha\mu}}{\partial x^{\nu}} = \Gamma_{\mu,\nu\alpha} + \Gamma_{\alpha,\mu\nu} \ .$$

$$-\frac{\partial g_{\nu\alpha}}{\partial x^{\mu}} = -\Gamma_{\alpha,\nu\mu} - \Gamma_{\nu,\alpha\mu} \ .$$

Taking half the sum of these equations and remembering that $\Gamma_{\mu,\nu\alpha} = \Gamma_{\mu,\alpha\nu}$, we find

$$\Gamma_{\mu,\nu\alpha} = \frac{1}{2}\left(\frac{\partial g_{\mu\nu}}{\partial x^{\alpha}} + \frac{\partial g_{\mu\alpha}}{\partial x^{\nu}} - \frac{\partial g_{\nu\alpha}}{\partial x^{\mu}}\right) . \quad (41)$$

From this we have for the symbols $\Gamma^{\mu}_{\nu\alpha} = g^{\mu\beta}\Gamma_{\beta,\nu\alpha}$

$$\Gamma^{\mu}{}_{\nu\alpha} = \frac{1}{2} g^{\mu\beta}\left(\frac{\partial g_{\beta\nu}}{\partial x^{\alpha}} + \frac{\partial g_{\beta\alpha}}{\partial x^{\nu}} - \frac{\partial g_{\nu\alpha}}{\partial x^{\beta}}\right) . \quad (42)$$

A coordinate system in which the Christoffel symbols vanish at point P is called a **geodesic coordinate system**, and the point P is said to be the pole.

4-5 Geodesics

As an application of the notion of parallel displacement and covariant differentiation, we now consider the geodesic equation. A geodesic is the curve defined by the requirement that each element of it is a parallel displacement of the preceding element. We shall see later that the world line of a point-like particle not acted upon by any forces, except gravitational, is a time-like geodesic.

If we take a point with coordinates x^{μ} and move it along a path, we then have x^{μ} as a function of some parameter s. There is a tangent vector $t^{\mu} = dx^{\mu}/ds$ at each point of the path. As we go along the path the vector t^{μ} gets shifted by parallel displacement: we shift the initial position from x^{μ} to $x^{\mu} + t^{\mu}ds$, and then shift the vector t^{μ} to this new position by parallel displacement, then shift the point again in the direction fixed by the new t^{μ}, and so on. If we are given the initial point and the initial value of the vector t^{μ}, not only the path can be determined, but also the parameter s along it. A path produced in this way is called a geodesic. We get the geodesic equations by applying Eq.(22) with $A^{\mu} = t^{\mu}$:

$$\frac{dt^{\nu}}{ds} + \Gamma^{\nu}{}_{\mu\sigma} t^{\mu} \frac{dx^{\sigma}}{ds} = 0 , \quad (43)$$

or

$$\frac{d^2x^\nu}{ds^2} + \Gamma^\nu_{\mu\sigma}\frac{dx^\sigma}{ds}\frac{dx^\mu}{ds} = 0 \ . \tag{44}$$

If the vector t^μ is initially a null vector, it always remains a null vector and the path is called a null geodesic. If the vector t^μ is initially time-like (i.e., $t^\mu t^\mu > 0$), it is always time-like and we have a time-like geodesic, and if t^μ is initially space-like ($t^\mu t^\mu < 0$), it is always space-like and we have a space-like geodesic.

For a time-like geodesic we may multiply the initial t^μ by a factor so as to make its length unity. This only requires a change in the scale of s. The vector t^μ now always has a unit length. It is simply the velocity vector $u^\mu = dx^\mu/d\tau$, and the parameter s has become the proper time τ. Eq.(43) becomes

$$\frac{du^\mu}{d\tau} + \Gamma^\nu_{\mu\sigma}u^\mu u^\sigma = 0 \tag{43a}$$

and Eq.(44) becomes

$$\frac{d^2x^\nu}{d\tau^2} + \Gamma^\nu_{\mu\sigma}\frac{dx^\sigma}{d\tau}\frac{dx^\mu}{d\tau} = 0 \tag{44a}$$

We make the physical assumption that the world line of a particle not acted on by any forces, except gravitational, is a time-like geodesic. Notice that the terms $-m\Gamma^\nu_{\mu\sigma}u^\mu u^\sigma$ in Eq.(43a) may be interpreted as the gravitational forces, and then the components of the metric tensor $g_{\mu\nu}$ play the role of the classical gravitational potential (as the Christoffel symbol is proportional to the derivatives of the metric tensor, see Eqs.(39) and (40)).

Choosing a local Galilean frame in which $g_{\mu\nu}$ = constants, the Christoffel symbols $\Gamma^\nu_{\mu\sigma} = 0$, and $du^\mu/d\tau = 0$. Therefore the gravitational forces can be locally eliminated, and the geodesic equations can be locally reduced to the special relativistic equations of motion, in agreement with the equivalence.

The path of a light ray is a null geodesic. It is fixed by Eq.(44) referring to some parameter s along the path. The proper time τ cannot now be used because $d\tau$ vanishes.

4-6 The stationary property of geodesics

We now examine the stationary property of geodesics. A geodesic that is not a null geodesic joining two points P and Q has a stationary value compared with the interval (line element) measured along another neighboring curve joining P and Q. This property holds good for a straight line in flat space and, in that case, it is also true that the straight line gives the shortest interval from one point to another. In curved space, the geodesic is no longer a straight line because space-time is no longer flat, and the particle motion is not rectilinear and uniform, in general. However, we can show that the geodesic is a path of extreme length. (We will not enquire whether or not the geodesic in a curved space gives the minimum or maximum value of the interval between any of its points). To show this, we demonstrate that the relations which must be satisfied to give a stationary value to the integral

$$s = \int \sqrt{g_{\lambda\mu} dx^{\lambda} dx^{\mu}} \tag{45}$$

are simply the equations of geodesic (44) of the previous section. Let us first introduce a parameter α and write Eq.(45) as

$$s = \int_{\alpha_p}^{\alpha_q} \left[g_{\lambda\mu} \frac{dx^{\lambda}}{d\alpha} \frac{dx^{\mu}}{d\alpha} \right]^{1/2} d\alpha$$

where α varies from point to point of the "geodesic" curve described by the relations which we are seeking for, and we write it as

$$x^{\mu} = f^{\mu}(\alpha) \ . \tag{46}$$

Any other neighboring curve joining P and Q has equations of the form

$$\bar{x}^{\lambda} = x^{\lambda} + \varepsilon y^{\lambda} = f^{\lambda}(\alpha) + \varepsilon y^{\lambda}(\alpha)$$

where $y^{\lambda} = 0$ at the end points P and Q, i.e., at $\alpha = \alpha_p$ and $\alpha = \alpha_q$, and ε is a small quantity whose square and higher powers are negligibly small. If s is the line element along the neighboring curve joining P and Q, then

$$\bar{s} = \int_{\alpha_p}^{\alpha_q} \sqrt{g_{\lambda\mu}(\bar{x}) \frac{d\bar{x}^{\lambda}}{d\alpha} \frac{d\bar{x}^{\mu}}{d\alpha}} d\alpha$$

and therefore, neglecting all powers of ε higher than the first,

$$\bar{s} - s = \int_{\alpha_p}^{\alpha_q} \sqrt{g_{\lambda\mu}\frac{dx^{\lambda}}{d\alpha}\frac{dx^{\mu}}{d\alpha} + \varepsilon\left(\frac{\partial g_{\lambda\mu}}{dx^{\sigma}}\frac{dx^{\alpha}}{d\lambda}y^{\sigma} + 2g_{\lambda\mu}\frac{dy^{\lambda}}{d\alpha}\right)\frac{dx^{\mu}}{d\alpha}}\, d\alpha - \int_{\alpha_p}^{\alpha_q}\sqrt{g_{\lambda\mu}\frac{dx^{\lambda}}{d\alpha}\frac{dx^{\mu}}{d\alpha}}\, d\alpha$$

which can be reduced to

$$\bar{s} - s = \frac{1}{2}\varepsilon\int_{\alpha_p}^{\alpha_q}\left[(y^{\sigma}\frac{\partial g_{\lambda\mu}}{\partial x^{\sigma}}\frac{dx^{\lambda}}{d\alpha} + 2g_{\lambda\mu}\frac{dy^{\lambda}}{d\alpha}\right]\frac{dx^{\mu}}{d\alpha}\frac{d\alpha}{ds}ds \ .$$

We can simplify the calculation, if $s \neq 0$, by assuming that the parameter α is identical with s itself measured along the "geodesic". For this case $d\alpha/ds = 1$ and

$$\bar{s} - s = \frac{1}{2}\int_{s_p}^{s_q}\left[y^{\sigma}\frac{\partial g_{\lambda\mu}}{dx^{\sigma}}\frac{dx^{\lambda}}{ds} + 2g_{\lambda\mu}\frac{dy^{\lambda}}{ds}\right]\frac{dx^{\mu}}{ds}ds$$

the x^{λ}, y^{λ} now being regarded as functions of s. Integration of the second term in the last equation by parts gives

$$\bar{s} - s = \frac{1}{2}\varepsilon\int_{s_p}^{s_q}\left[\frac{\partial g_{\lambda\mu}}{dx^{\sigma}}\frac{dx^{\lambda}}{ds}\frac{dx^{\mu}}{ds} - 2\frac{d}{ds}\left(g_{\sigma\mu}\frac{dx^{\mu}}{ds}\right)\right]y^{\sigma}ds + \varepsilon\left[g_{\lambda\mu}\frac{dx^{\mu}}{ds}y^{\lambda}\right]_{s_p}^{s_q} .$$

But the functions y^{λ} vanish at $s = s_p$ and $s = s_q$, and therefore the integrated term is zero. Therefore, if the interval is to have a stationary value for the "geodesic" curve compared with neighboring curves, $\bar{s} - s$ must be zero for any choice of the function y^{λ}. This is possible only if the coefficient of each y^{σ} in the integrand is separately zero, and therefore the differential equations of the "geodesic" are the n equations

$$\frac{d}{ds}\left(g_{\sigma\mu}\frac{dx^{\mu}}{ds}\right) - \frac{1}{2}\frac{\partial g_{\lambda\mu}}{\partial x^{\sigma}}\frac{dx^{\lambda}}{ds}\frac{dx^{\mu}}{ds} = 0, \quad (\sigma = 1,2,...n) \ . \qquad (47)$$

Now
$$\frac{d}{ds}\left(g_{\sigma\mu}\frac{dx^{\mu}}{ds}\right) = g_{\sigma\mu}\frac{d^2x^{\mu}}{ds^2} + \frac{\partial g_{\mu\sigma}}{\partial x^{\lambda}}\frac{dx^{\lambda}}{ds}\frac{dx^{\mu}}{ds}$$

$$= g_{\sigma\mu}\frac{d^2x^{\mu}}{ds^2} + \frac{1}{2}(g_{\mu\sigma,\lambda} + g_{\lambda\sigma,\mu})\frac{dx^{\lambda}}{ds}\frac{dx^{\mu}}{ds} .$$

Thus the equations of the "geodesic" may be written

$$g_{\sigma\mu}\frac{d^2x^{\mu}}{ds^2} + \frac{1}{2}(g_{\mu\sigma,\lambda} + g_{\lambda\sigma,\mu} - g_{\lambda\mu,\sigma})\frac{dx^{\lambda}}{ds}\frac{dx^{\mu}}{ds}$$

$$= g_{\sigma\mu}\frac{d^2x^{\mu}}{ds^2} + \Gamma_{\sigma,\lambda\mu}\frac{dx^{\lambda}}{ds}\frac{dx^{\mu}}{ds} = 0 \cdot \qquad (48)$$

Multiplying this by $g^{\tau\sigma}$ and summing over σ, the result is

$$\frac{d^2x^{\tau}}{ds^2} + \Gamma^{\tau}_{\lambda\mu}\frac{dx^{\lambda}}{ds}\frac{dx^{\mu}}{ds} = 0, \qquad (\tau = 1,2,...n) \qquad (49)$$

which is simply the standard form of Eq.(44) for geodesics.

The above work shows that we may use the stationary condition as the definition of a geodesic, except in dealing with the propagation of light. In that case, we have null geodesics, and the deduction as given above cannot be applied because ds vanishes throughout.

4-7 The curvature tensor

In a flat space, if we perform two (ordinary) differentiations in succession their order does not matter. However, this does not, in general, hold for covariant differentiation in a curved space, except for a scalar ϕ. For the case of a scalar, we have

$$\phi_{;\mu;\nu} = (\phi_{;\mu})_{,\nu} - \Gamma^{\alpha}_{\mu\nu}\phi_{;\alpha} = \phi_{,\mu,\nu} - \Gamma^{\alpha}_{\mu\nu}\phi_{,\alpha} \cdot \qquad (50)$$

Since $\Gamma^{\alpha}_{\mu\nu}$ is symmetric in the lower indexes μ and ν, so in this case the order of differentiation does not matter.

Now if we take a contravariant vector A^{μ} and apply covariant differentiations twice to it, we will find the order of differentiation is very important. First, the covariant differentiation of A^{μ} gives a mixed tensor

$$A^{\mu}_{;\nu} = \frac{\partial A^{\mu}}{\partial x^{\nu}} + \Gamma^{\alpha}_{\alpha\nu} A^{\alpha} .$$

Covariant differentiation of this mixed tensor gives

$$A^{\mu}_{;\nu;\beta} = \frac{\partial}{\partial x^{\beta}}(A^{\mu}_{;\nu}) + \Gamma^{\mu}_{\alpha\beta} A^{\alpha}_{;\nu} - \Gamma^{\alpha}_{\nu\beta} A^{\mu}_{;\alpha}$$

or

$$A^{\mu}{}_{;\nu;\beta} = \frac{\partial^2 A^{\mu}}{\partial x^{\beta}\partial x^{\nu}} + \Gamma^{\mu}{}_{\alpha\nu}\frac{\partial A^{\alpha}}{\partial x^{\beta}} + A^{\alpha}\frac{\partial \Gamma^{\mu}{}_{\alpha\nu}}{\partial x^{\beta}} + A^{\alpha}{}_{;\nu}\Gamma^{\mu}{}_{\alpha\beta} - A^{\mu}{}_{;\alpha}\Gamma^{\alpha}{}_{\nu\beta} . \tag{51}$$

Interchanging β and ν, we obtain

$$A^{\mu}{}_{;\beta;\nu} = \frac{\partial^2 A^{\mu}}{\partial x^{\nu}\partial x^{\beta}} + \Gamma^{\mu}{}_{\alpha\beta}\frac{\partial A^{\alpha}}{\partial x^{\nu}} + A^{\alpha}\frac{\partial \Gamma^{\mu}{}_{\alpha\beta}}{\partial x^{\nu}} + A^{\alpha}{}_{;\beta}\Gamma^{\mu}{}_{\alpha\nu} - A^{\mu}{}_{;\alpha}\Gamma^{\alpha}{}_{\beta\nu} . \tag{52}$$

Subtracting Eq.(52) from Eq.(51), we obtain

$$A^{\mu}{}_{;\nu;\beta} - A^{\mu}{}_{;\beta;\nu} = A^{\alpha}\left[\frac{\partial \Gamma^{\mu}_{\alpha\nu}}{\partial x^{\beta}} - \frac{\partial \Gamma^{\mu}_{\alpha\beta}}{\partial x^{\nu}} + \Gamma^{\gamma}_{\alpha\nu}\Gamma^{\mu}_{\gamma\beta} - \Gamma^{\gamma}_{\alpha\beta}\Gamma^{\mu}_{\gamma\nu}\right] \tag{53}$$

$$= A^{\alpha} R^{\mu}_{\alpha\nu\beta}$$

where

$$R^{\alpha}{}_{\beta\gamma\delta} = \frac{\partial \Gamma^{\alpha}_{\beta\gamma}}{\partial x^{\delta}} - \frac{\partial \Gamma^{\alpha}_{\beta\delta}}{\partial x^{\gamma}} + \Gamma^{\mu}_{\beta\gamma}\Gamma^{\alpha}_{\mu\delta} - \Gamma^{\mu}_{\beta\delta}\Gamma^{\alpha}_{\mu\gamma} . \tag{54}$$

Since $A^{\mu}_{;\nu;\beta} - A^{\mu}_{;\beta;\nu}$ and A^{α} are tensors, $R^{\mu}_{\alpha\nu\beta}$ must be the component of a tensor, by the quotient law. It is called the curvature tensor or the Riemann-Christoffel tensor, and it depends only on the Christoffel symbols and their derivatives. In a flat space all $g_{\mu\nu}$ can be transformed into constants (rectangular or Galilean coordinates) where all Christoffel symbols vanish; hence $R^{\mu}_{\alpha\nu\beta} = 0$. Being a tensor equation, it holds in all coordinate systems (Cartesian, oblique, or curvilinear). In a curved space, $R^{\mu}_{\alpha\nu\beta}$ will not vanish. Therefore it is a measure of the curvature of space.

From the expression (54) it follows that the curvature tensor is antisymmetric in the indices γ and δ:

$$R^{\alpha}{}_{\beta\gamma\delta} = -R^{\alpha}{}_{\beta\delta\gamma} \ . \tag{55}$$

Furthermore, it is easy to verify that the following identity is valid:

$$R^{\alpha}{}_{\beta\gamma\delta} + R^{\alpha}{}_{\delta\beta\gamma} + R^{\alpha}{}_{\gamma\delta\beta} = 0 \ . \tag{56}$$

In addition to the mixed curvature tensor $R^{\alpha}{}_{\beta\gamma\delta}$, we can also use the covariant curvature tensor

$$R_{\alpha\beta\gamma\delta} = g_{\alpha\eta} R^{\eta}{}_{\beta\gamma\delta} \ . \tag{57}$$

From the transformation law for $R_{\alpha\beta\gamma\delta}$

$$R_{\alpha\beta\gamma\delta} = \frac{1}{2}\left(\frac{\partial^2 g_{\alpha\delta}}{\partial x^{\beta} \partial x^{\gamma}} + \frac{\partial^2 g_{\beta\gamma}}{\partial x^{\alpha} \partial x^{\delta}} - \frac{\partial^2 g_{\alpha\gamma}}{\partial x^{\beta} \partial x^{\delta}} - \frac{\partial^2 g_{\beta\delta}}{\partial x^{\alpha} \partial x^{\gamma}} \right) + g_{\mu\nu}\left(\Gamma^{\mu}_{\beta\gamma}\Gamma^{\nu}_{\alpha\delta} - \Gamma^{\mu}_{\beta\delta}\Gamma^{\nu}_{\alpha\gamma}\right) \tag{58}$$

(It is not difficult to derive this transformation law, but it is tedious. We simply give it here, without derivation.) We can see the following symmetry properties:

$$R_{\alpha\beta\gamma\delta} = -R_{\beta\alpha\gamma\delta} = -R_{\alpha\beta\delta\gamma}, \quad R_{\alpha\beta\gamma\delta} = R_{\gamma\delta\alpha\beta} \tag{59}$$

i.e., the tensor is antisymmetric in each of the index pairs α, β and γ, δ and is symmetric under the interchange of the two pairs with each other. Thus, all components $R_{\alpha\beta\gamma\delta}$, in which $\alpha = \beta$ or $\gamma = \delta$ are zero. As for $R^{\alpha}{}_{\beta\gamma\delta}$, we also have the identity

$$R_{\alpha\beta\gamma\delta} + R_{\alpha\delta\beta\gamma} + R_{\alpha\gamma\delta\beta} = 0 \ . \tag{60}$$

A 4th-rank tensor has $4^4 = 256$ components. However, because of the above symmetries, the number of algebraically independent components of $R_{\alpha\beta\gamma\delta}$ is only 20.

By contracting the curvature tensor, we get the symmetric tensor, called the Ricci tensor:

$$R_{\sigma\mu} = R^{\lambda}{}_{\sigma\mu\lambda} = R_{\mu\sigma} \,. \tag{61}$$

According to Eq.(54), we have

$$R_{\sigma\mu} = \frac{\partial\Gamma^{\lambda}{}_{\sigma\mu}}{\partial x^{\lambda}} - \frac{\partial\Gamma^{\lambda}{}_{\sigma\lambda}}{\partial x^{\mu}} + \Gamma^{\lambda}{}_{\sigma\mu}\Gamma^{\nu}{}_{\lambda\nu} - \Gamma^{\nu}{}_{\sigma\lambda}\Gamma^{\lambda}{}_{\mu\nu} \,. \tag{62}$$

This tensor is clearly symmetric: $R_{\sigma\mu} = R_{\mu\sigma}$.

Finally, contracting $R_{\sigma\mu}$, we obtain

$$R = g^{\sigma\mu}R_{\sigma\mu} = g^{\sigma\lambda}g^{\mu\beta}R_{\sigma\mu\lambda\beta} \tag{63}$$

which is called the scalar curvature of the space.

One of the most important tensors in the study of gravitation is the Einstein tensor, defined by

$$G_{\mu\nu} = R_{\mu\nu} - \frac{1}{2}g_{\mu\nu}R = G_{\nu\mu} \tag{64}$$

The Einstein tensor $G_{\mu\nu}$ is not unique; we can add any tensor whose divergence vanishes to $G_{\mu\nu}$ without changing it. For example, $\Lambda g_{\mu\nu}$ where Λ is the so-called cosmological constant ($\sim 10^{-56}$ cm^{-2}), can be added to $G_{\mu\nu}$ without changing it.

4-8 The condition for flat space

If space is flat, we may choose a system of coordinates that is rectilinear, and then the $g_{\mu\nu}$ are all constant. The tensor $R_{\mu\nu\rho\sigma}$ then vanishes.

Conversely, if $R_{\mu\nu\rho\sigma}$ vanishes, one can prove that the space is flat.

4-9 Geodesic deviation

We can get a good insight into the nature of a space just by examining the problem of geodesic deviation. For example, consider two nearby freely falling particles which travel on paths $x^\mu(\tau)$ and $x^\mu(\tau) + \delta x^\mu(\tau)$. The equations of motion are then given by

$$\frac{d^2x^\mu}{d\tau^2} + \Gamma^\mu{}_{\nu\lambda}(x)\frac{dx^\nu}{d\tau}\frac{dx^\lambda}{d\tau} = 0 \tag{65}$$

and

$$\frac{d^2}{d\tau^2}\left[x^\mu + \delta x^\mu\right] + \Gamma^\mu{}_{\nu\lambda}(x + \delta x)\frac{d}{d\tau}\left[x^\nu + \delta x\right]\frac{d}{d\tau}\left[x^\lambda + \delta x^\lambda\right] = 0\ . \tag{66}$$

Evaluating the difference between these equations to first order in δx^μ gives

$$\frac{d^2\delta x^\mu}{d\tau^2} + \frac{\partial\Gamma^\mu{}_{\nu\lambda}}{\partial x_\rho}\delta x^\rho\frac{dx^\nu}{d\tau}\frac{dx^\lambda}{d\tau} + 2\Gamma^\mu{}_{\nu\lambda}\frac{dx^\nu}{d\tau}\frac{d\delta x^\lambda}{d\tau} = 0 \tag{67}$$

or, in terms of covariant derivatives along the curves $x^\mu(\tau)$

$$\frac{D^2}{D\tau^2}\delta x^\lambda = R^\lambda{}_{\nu\mu\rho}\delta x^\mu\frac{dx^\nu}{d\tau}\frac{dx^\rho}{d\tau}\ . \tag{68}$$

Eq.(68) is called the equation of geodesic deviation.

Although a freely falling particle appears to be at rest in a coordinate system falling with the particle, a pair of nearby freely falling particles will exhibit a relative motion which can reveal the presence of a gravitational field to an observer who falls with the particles. The effect of the right hand side of Eq.(68) becomes negligible when the separation between particles is much less than the characteristic dimensions of the field. This indicates clearly that the **local inertial frames** are only locally applicable, otherwise, the principle of equivalence will be violated.

4-10 Laws of physics in curved spaces

The laws of physics must be valid in all coordinate systems. They must thus be expressible as tensor equations. Whenever they involve the derivative of a field quantity, it must be a covariant derivative. The field equations of physics must be rewritten with the ordinary derivatives replaced by covariant derivatives.

Even if we are working with flat space (which means neglecting the gravitational field) and we are using curvilinear coordinates, we must write our equations in terms of covariant derivatives if we want them to hold in all coordinate systems.

As for the problem of generalizing a particular physics law from the flat Minkowski space to a general curved space, there is no unique solution at all. In fact, the problem is highly complicated since, in general, there will be an interaction between the space (the $g_{\mu\nu}$) and the physical phenomenon whose laws we are trying to formulate. But if the object under consideration does not appreciably influence the $g_{\mu\nu}$, that is, if the $g_{\mu\nu}$ are determined by objects much more massive than the object under consideration, we may then consider the $g_{\mu\nu}$ as given functions of the space-time variables, $g_{\mu\nu}(x^\sigma)$. In this case the geometry is rigidly determined and the effect of the physical object under study on the geometrical structure may be neglected. Under these circumstances, we may take over the special relativistic laws by substituting

$$d \longrightarrow D, \qquad \partial\mu \longrightarrow D\mu, \qquad d\Omega \longrightarrow \sqrt{-g}\,d\Omega \ .$$

where

$$\partial_\mu A^\nu = \frac{\partial A^\nu}{\partial x^\mu}, \qquad D_\mu A = A^\mu_{;\nu} \ .$$

As an example, consider the motion of a free particle. Its time track in Minkowski space is characterized by the equations

$$d^2x^i/ds^2 = 0, \quad i = 0,1,2,3 \ . \tag{69}$$

These equations imply a straight line in the four-dimensional Minkowski space, which in turn corresponds to a uniform rectilinear motion in three-dimensional space. These equations can be derived from the stationary condition that the integral $\int ds$, taken along the motion between two points P and Q, is stationary if one makes a small variation of the path keeping the end-points fixed:

$$\delta \int_P^Q ds = 0 \tag{70}$$

where the variation vanishes at the end-points P and Q.

If the particle is subject to the action of a gravitational field, its equation of motion is no longer a straight line, because the space-time is curved. But the particle still follows a stationary trajectory, the geodesic. As shown above, the geodesic equation can be obtained from the same variational principle (70), provided that the Minkowski metric is replaced by the curved space metric $g_{\mu\nu}$ and $ds^2 = g_{\mu\nu}dx^\mu dx^\nu$. Instead of computing explicitly the variation in Eq.(70), we can simply obtain the geodesic equations as the covariant generalization of Eq.(69):

$$D^2x^i/Ds^2 = 0$$

which is equivalent to

$$d^2x^i/ds^2 + \Gamma^\alpha_{\mu\nu}(dx^\mu/ds)(dx^\nu/ds) = 0,$$

the geodesic equations.

4-11 The metric tensor and the classical gravitational potential

The presence of a gravitational field modifies the structure of space-time. Any gravitational field is just a change in the metric of space-time, as determined by the metric tensor $g_{\mu\nu}$. Through the geodesic equations of motion, we can now provide the expressions governing the union of geometry and gravitation. To this end, let us compare the Newtonian equation of motion of a particle in a gravitational field and its geodesic equations of motion in a curved-space geometry:

$$d^2x^\alpha/ds^2 + \partial\phi/\partial x^\alpha = 0 \tag{71}$$

$$d^2x^\alpha/ds^2 + \Gamma^\alpha_{\mu\nu}(dx^\mu/ds)(dx^\nu/ds) = 0 \tag{72}$$

where ϕ is the Newtonian gravitational potential. These two equations have a fundamental similarity in that both are independent of the mass of the moving body under consideration. Thus, both equations satisfy the principle of equivalence. Now since the derivative d^2x^α/ds^2 is the four-acceleration of the particle, the quantity $-m\Gamma^\alpha_{\mu\nu}u^\mu u^\nu$ may be interpreted as the gravitational force, and then the components of the metric tensor $g_{\mu\nu}$ play the role of the Newtonian gravitational potential ϕ (as the Christoffel symbols are constructed from the derivatives of the $g_{\mu\nu}$). We must first show that this interpretation is consistent with the

Newtonian equations of motion; namely, we must show that in the limit of ordinary velocities the geodesic equations reduce to the Newtonian equations. To see this, let the velocity $dx^\alpha/dt << c$. Then $ds^2 \simeq g_{00}c^2dt^2$, and $ds \simeq \sqrt{g_{00}}\, c\, dt$, so that

$$\simeq d^2x^\alpha/dt^2 + \Gamma^\alpha_{00}\, c^2 = 0,$$

where

$$\Gamma^\alpha_{00} = \partial g_{00}/\partial x^\alpha$$

From this we see that in this limit $g_{00} = K + 2\phi/c^2$. Since in flat space $g_{00} = 1$, we have $K = 1$ and

$$g_{00} = 1 + 2\phi/c^2 \; . \tag{73}$$

This shows that the identification postulated above is plausible, i.e., the metric tensor $g_{\mu\nu}$ plays the role of Newtonian gravitational potential.

We should be careful to note that the physical content of the two equations (equations (71) and (72)) are entirely different. In Newton's equation we have a field ϕ, which causes the motion. The particle is under a force and its velocity changes in time. In the geodesic equations, on the other hand, there is no physical agent such as ϕ. The particle follows a geodesic which is determined by the geometry of the space-time. This change in interpretation is actually a conceptual simplification, since inertia and gravitation are unified and the concept of external force is eliminated from the theory of gravitation.

4-12 Some useful calculational aids

We conclude this chapter by giving a number of useful aids in manipulation of tensor quantities.

First show that the covariant derivative of the metric tensor $g_{\mu\nu}$ is zero. To do this we note that the relation

$$DA_\mu = g_{\mu\nu} DA^\nu \tag{74}$$

is valid for the vector DA_μ, as for any vector. On the other hand,

$$A_\mu = g_{\mu\nu} A^\nu$$

so that

$$DA_\mu = D(g_{\mu\nu}A^\nu) = g_{\mu\nu}DA^\nu + A^\nu Dg_{\mu\nu} .$$

Comparing this with Eq.(74), and remembering that the vector A^ν is arbitrary, we have

$$Dg_{\mu\nu} = 0 .$$

Therefore the covariant derivative of $g_{\mu\nu}$ is zero:

$$g_{\mu\nu;\lambda} = 0 . \tag{75}$$

Thus $g_{\mu\nu}$ may be considered as a constant during covariant differentiation.

We next derive an expression for the contracted Christoffel symbol $\Gamma^\mu_{\alpha\mu}$ which will be very useful later on. From Eq.(39) we have

$$\Gamma^\mu_{\alpha\mu} = \frac{1}{2}g^{\mu\nu}\left(\frac{\partial g_{\nu\alpha}}{\partial x^\mu} + \frac{\partial g_{\nu\mu}}{\partial x^\alpha} - \frac{\partial g_{\alpha\mu}}{\partial x^\nu}\right) .$$

Changing the positions of the dummy indexes μ and ν in the first term and remembering $g_{\mu\nu} = g_{\nu\mu}$, we see that the first and third terms then cancel each other, so that

$$\Gamma^\mu_{\alpha\mu} = \frac{1}{2}g^{\mu\nu}\frac{\partial g_{\mu\nu}}{\partial x^\alpha} \tag{76}$$

which can be simplified. To do this we calculate the differential dg of the determinant g made up from the components of the metric tensor $g_{\mu\nu}$; dg can be obtained by taking the differential of each component of the tensor $g_{\mu\nu}$ and multiplying it by its coefficient in the determinant, i.e. by the corresponding minor:

$$dg = dg_{\mu\nu}M^{\mu\nu}$$

where $M^{\mu\nu}$ is the minor of the component $g_{\mu\nu}$. Now

$$g^{\mu\nu} = \frac{M^{\mu\nu}}{g} , \qquad M^{\mu\nu} = g^{\mu\nu}g .$$

Thus

$$dg = g\,g^{\mu\nu}\,dg_{\mu\nu} = -g\,g_{\mu\nu}dg^{\mu\nu}\ .$$

The expression on the right of the above equation follows from

$$d\left(g_{\mu\nu}g^{\mu\nu}\right) = d\left(\delta^{\mu}_{\mu}\right) = d(4) = 0$$

We then have

$$\frac{\partial g}{\partial x^{\alpha}} = g\,g^{\mu\nu}\frac{\partial g_{\mu\nu}}{\partial x^{\alpha}} = -g\,g_{\mu\nu}\frac{\partial g^{\mu\nu}}{\partial x^{\alpha}} \tag{77}$$

The use of Eq.(77) enables us to write Eq.(76) in the form

$$\Gamma^{\mu}_{\alpha\mu} = \frac{1}{2}g^{\mu\nu}\frac{\partial g_{\mu\nu}}{\partial x^{\alpha}} = \frac{1}{2g}\frac{\partial g}{\partial x^{\alpha}} = \frac{\partial \ln\sqrt{-g}}{\partial x^{\alpha}} \tag{78}$$

This expression is very useful. First consider the covariant divergence $A^{\mu}_{;\mu}$:

$$A^{\mu}_{;\mu} = \frac{\partial A^{\mu}}{\partial x^{\mu}} + \Gamma^{\mu}_{\alpha\mu}A^{\alpha}\ .$$

Substituting the expression (78) for $\Gamma^{\mu}_{\alpha\mu}$, we obtain

$$A^{\mu}_{;\mu} = \frac{\partial A^{\mu}}{\partial x^{\mu}} + A^{\alpha}\frac{\partial \ln\sqrt{-g}}{\partial x^{\alpha}} = \frac{1}{\sqrt{-g}}\frac{\partial\left(\sqrt{-g}\,A^{\mu}\right)}{\partial x^{\mu}}\ . \tag{79}$$

We now consider the covariant divergence of a contravariant tensor of the second rank $T^{\mu\nu}_{;\alpha}$. From Eq.(27), we have

$$T^{\mu\nu}_{;\alpha} = \frac{\partial T^{\mu\nu}}{\partial x^{\nu}} + \Gamma^{\mu}_{\beta\nu}T^{\beta\nu} + \Gamma^{\mu}_{\beta\alpha}T^{\mu\beta}\ .$$

Changing the positions of the dummy indexes μ and β in the third term on the right-hand side, we obtain

$$T^{\mu\nu}_{;\nu} = \frac{\partial T^{\mu\nu}}{\partial x^{\nu}} + \Gamma^{\mu}_{\beta\nu}T^{\beta\nu} + \Gamma^{\beta}_{\nu\beta}T^{\mu\nu}\ .$$

Substituting the expression (78) for $\Gamma^{\beta}_{\nu\beta}$, we obtain

$$T^{\mu\nu}_{;\nu} = \frac{\partial T^{\mu\nu}}{\partial x^{\nu}} + \Gamma^{\mu}_{\beta\nu}T^{\beta\nu} + \frac{\partial \ln\sqrt{-g}}{\partial x^{\nu}}T^{\mu\nu}$$

or, finally

$$T^{\mu\nu}_{;\nu} = \frac{1}{\sqrt{-g}}\frac{\partial}{\partial x^{\nu}}\left(\sqrt{-g}\,T^{\mu\nu}\right) + \Gamma^{\mu}_{\beta\nu}T^{\beta\nu} \ . \tag{80}$$

Similarly, for a mixed tensor, Eq.(28) leads to

$$T^{\beta}_{\alpha;\beta} = \frac{1}{\sqrt{-g}}\frac{\partial}{\partial x^{\beta}}\left(T^{\beta}_{\alpha}\sqrt{-g}\right) - \Gamma^{\beta}_{\alpha\sigma}T^{\sigma}_{\ \beta} \ . \tag{81}$$

For an antisymmetric tensor $F^{\beta\nu} = -F^{\nu\beta}$, then

$$\Gamma^{\mu}_{\beta\nu}F^{\beta\nu} = \Gamma^{\mu}_{\nu\beta}F^{\nu\beta} = -\Gamma^{\mu}_{\beta\nu}F^{\beta\mu} \ . \tag{82}$$

The expression in the middle is obtained by interchanging the dummy indexes β and ν, and the expression on the far right follows from the $F^{\beta\nu} = -F^{\nu\beta}$ and $\Gamma^{\mu}_{\nu\beta} = \Gamma^{\mu}_{\beta\nu}$. From Eq.(82) it follows that

$$\Gamma^{\mu}_{\beta\nu}F^{\beta\mu} = 0 \ .$$

Thus, for an antisymmetric tensor $F^{\alpha\beta}$, the last term of Eq.(80) vanishes and the covariant divergence is

$$F^{\alpha\beta}_{\ ;\beta} = \frac{1}{\sqrt{-g}}\frac{\partial}{\partial x^{\beta}}\left(F^{\alpha\beta}\sqrt{-g}\right) \ . \tag{83}$$

For a symmetric tensor $S^{\alpha\beta}$, rearrangement of the last term of Eq.(81) leads to

$$S^{\beta}_{\alpha;\beta} = \frac{1}{\sqrt{-g}}\frac{\partial}{\partial x^{\beta}}\left(S^{\beta}_{\alpha}\sqrt{-g}\right) - \frac{1}{2}\frac{\partial g_{\beta\nu}}{\partial x^{\alpha}}S^{\beta\nu} \ . \tag{84}$$

In Cartesian coordinates, the curl

$$\frac{\partial A_{\alpha}}{\partial x^{\beta}} - \frac{\partial A_{\beta}}{\partial x^{\alpha}}$$

is an antisymmetric tensor. In curvilinear coordinates this tensor is

$$A_{\alpha;\beta} - A_{\beta;\alpha}$$

However, with the help of the expression for $A_{\alpha;\beta}$ and since $\Gamma^{\alpha}_{\beta\gamma} = \Gamma^{\alpha}_{\gamma\beta}$, we have

$$A_{\alpha;\beta} - A_{\beta;\alpha} = \frac{\partial A_\alpha}{\partial x^\beta} - \frac{\partial A_\beta}{\partial x^\alpha} \qquad (85)$$

i.e., the ordinary curl is still a tensor (antisymmetric) in curvilinear coordinates.

Finally, we transform to curvilinear coordinates the sum

$$\frac{\partial^2 \varphi}{\partial x_\alpha \partial x^\alpha}$$

of the second derivatives of a scalar φ. In curvilinear coordinates this sum goes over to $\varphi_{;\alpha}^{;\alpha}$. But covariant differentiation of a scalar reduces to ordinary differentiation:

$$\varphi_{;\alpha} = \frac{\partial \varphi}{\partial x^\alpha}$$

Raising the index α, we have

$$\varphi^{;\alpha} = g^{\alpha\beta}\frac{\partial \varphi}{\partial x^\beta}$$

which is a contravariant tensor of rank one. Using formula (79), we find

$$\varphi_{;\alpha}^{;\alpha} = \frac{1}{\sqrt{-g}}\frac{\partial}{\partial x^\alpha}\left(\sqrt{-g}\, g^{\alpha\beta}\frac{\partial \varphi}{\partial x^\beta}\right) . \qquad (86)$$

In view of Eq.(79), Gauss' theorem for transformatiol of the integral of a vector over a hypersurface into an integral over a four-volume can now be written as

$$\oint A^\alpha \sqrt{-g}\, dS_\alpha = \int A^\alpha_{;\alpha}\sqrt{-g}\, d\Omega . \qquad (87)$$

Problems

(4-1) Write the terms in each of the following indicated sums

a) $a_{jk}x^{k}$ b) $A_{pq}B^{qr}$ c) $\bar{g}_{rs} = g_{jk}\dfrac{\partial x^{j}}{\partial \bar{x}^{r}}\dfrac{\partial x^{k}}{\partial \bar{x}^{s}}$

(4-2) Write the transformation law for the following tensors

a) $A^{i}{}_{jk}$ b) $B^{pq}{}_{ijk}$

(4-3) A quantity A(j,k,m,n) which is a function of the coordinates x^{i} transforms to another coordinate system x^{i} according to the rule

$$\bar{A}(p,q,r,s) = \frac{\partial x^{j}}{\partial \bar{x}^{p}}\frac{\partial \bar{x}^{q}}{\partial x^{k}}\frac{\partial \bar{x}^{r}}{\partial x^{m}}\frac{\partial \bar{x}^{s}}{\partial x^{n}}A(j,k,m,n)$$

Is the quantity a tensor? If so, write the tensor in suitable notation and give the covariant and contravariant rank.

(4-4) Show that the property of symmetry (or antisymmetry) with respect to indexes of a tensor is invariant under coordinate transformation.

(4-5) A covariant tensor has components xy, $2y - z^{2}$, xz in rectangular coordinates; find its covariant components in spherical coordinates.

(4-6) Prove that the contraction of the outer product of the two tensors A^{p} and B_{p} is a scalar.

(4-7) Determine the Christoffel symbols of the second kind in rectangular and cylindrical coordinates.

(4-8) The line element on the surface of a sphere of radius a in Euclidean space is given by

$$ds^{2} = a^{2}(d\theta^{2} + \sin^{2}\theta d\phi^{2}) .$$

For this space calculate $\Gamma^{i}_{k\ell}$, i, k, $\ell = 1,2$ (with $\theta = x^{1}$, $\phi = x^{2}$).

(4-9) Find the covariant derivative of $A^{i}{}_{j}\ B^{km}{}_{n}$ with respect to x^{q}

(4-10) Prove that

$$\nabla \cdot A^{p} = \frac{1}{\sqrt{-g}} \frac{\partial}{\partial x^{k}} \left(\sqrt{-g} A^{k} \right)$$

and

$$\nabla^2 \phi = \frac{1}{\sqrt{-g}} \frac{\partial}{\partial x^{k}} \left(\sqrt{-g}\, g^{kr} \frac{\partial \phi}{\partial x^{r}} \right) .$$

(4-11) Determine the force acting on a particle in a constant gravitational field.

Problems

Chapter 5

Einstein's Law of Gravitation

5-1 Introduction (Summary of general principles)

We now summarize the concepts and ideas discussed so far as a series of postulates. They will indicate the correct approach for establishing the new laws of gravitation.

[1] Under a general coordinate transformation, the laws of physics remain covariant (i.e., form invariant). This requirement is known as the principle of general covariance.

[2] When gravitational fields are present, space-time is curved endowed with a metric of the form

$$ds^2 = g_{\mu\nu} \quad dx^\mu dx^\nu \qquad (\mu, \nu = 0,1,2,3) \ .$$

The metric tensor $g_{\mu\nu}$ are functions of the space-time variables x^0, x^1, x^2, x^3.

[3] The metric tensor $g_{\mu\nu}$ can be interpreted as the generalization of the gravitational field Φ. This is an indication of the union of gravitation and geometry.

[4] According to Mach's principle, the curvature of space-time is governed by the masses embedded in it.

[5] In the nonrelativistic limit, Newtonian dynamics and Newtonian gravitational theory are both valid. We can call this the correspondence principle.

According to [1], the new laws of gravitation must be expressed in tensorial form. While [2] and [3] suggest that the new laws at least contain 10 quantities which are related to $g_{\mu\nu}$, and [4] says that the energy-momentum tensor must be related to the curvature of space-time. Finally, [5] says that in the nonrelativistic limit, our new laws of gravitation reduce to Poisson's equation.

5-2 A heuristic derivation of Einstein's equations

It is not clear how one can formulate precisely the laws of gravitation starting from such extremely general postulates. It is Einstein's triumph and an example "of the power of speculative thought" that he was able to formulate the laws of gravitation starting from such general principles. We will follow Chandrasekhar's approach in formulating the laws of gravitation, namely, we ask three basic questions, and answer them first in terms of Newtonian theory, then via their relativisitic generalizations. The three basic questions are:

(1) When can we say that there is no gravitational field present?

(2) What are the equations which determine the gravitational field in empty space, outside material bodies?

(3) What are the equations in regions of space where matter is present? ("Empty" here means that there is no matter present and no physical fields except gravitational field.)

In the Newtonian theory, gravitational field is to be described in terms of a scalar potential function Φ; and the equations that determine Φ are

(i) $$\Phi = 0 \qquad (1)$$

when there is no gravitation;

(ii) $$\nabla^2\Phi = 0 \qquad \text{(Laplace equation)} \qquad (2)$$

in empty space; and

(iii) $$\nabla^2\Phi = 4\pi G\rho \quad \text{(Poisson's equation)} \tag{3}$$

in regions of space where matter is present and the material density is ρ; these equations are supplemented by the standard equations of motion.

A. Vacuum field equations

We first seek relativistic generalizations of (i) and (ii). The generalization of (iii) will be addressed later.

How can we generalize the condition $\Phi = 0$? Let us start with Newton's first law of motion, which states that in the absence of a gravitational field, a free particle can experience no acceleration, i.e., in a Cartesian frame

$$d^2x^i/ds^2 = 0\ . \tag{4}$$

If we should describe the motion in curvilinear coordinates, the particle will be subjected to inertial accelerations and the equation of motion of the particle is a geodesic equation

$$d^2x^i/ds^2 + \Gamma^i_{jk}(dx^j/ds)(dx^k/ds) = 0\ . \tag{5}$$

We may therefore conclude that no gravitational field is present if a coordinate transformation is possible such that in the new coordinate system the equations of motion reduce to Eq.(4). In other words, in the new coordinate system the Christoffel symbols Γ^i_{jk} all vanish identically. Then the curvature tensor $R^\alpha{}_{\beta\gamma\delta}$

$$R^\alpha{}_{\beta\gamma\delta} = \frac{\partial\Gamma^\alpha_{\beta\gamma}}{\partial x^\delta} - \frac{\partial\Gamma^\alpha_{\beta\gamma}}{\partial x^\gamma} + \Gamma^\mu{}_{\beta\gamma}\Gamma^\alpha{}_{\mu\delta} - \Gamma^\mu{}_{\beta\delta}\Gamma^\alpha{}_{\mu\gamma}$$

also vanishes identically. We may now state that the condition for the absence of any gravitational field is the vanishing of the Riemann Christoffel curvature tensor:

$$R^\alpha{}_{\beta\gamma\delta} = 0\ . \tag{6}$$

This corresponds to the requirement $\Phi = 0$ in the Newtonian theory.

We now need a generalization of Eq.(6), in the sense that in the Newtonian theory $\nabla^2\Phi = 0$ is a generalization of $\Phi = 0$. Now, the gravitational field intensity $\vec{g}$ is derivable from a potential Φ in the usual way $\vec{g} = -\nabla\Phi$, or

$$g_i = -\frac{\partial\Phi}{\partial x^i} = -\Phi_{,i} \qquad (i = 1,2,3) . \tag{7}$$

Now considering a pair of freely falling particles separated by a small separation ζ^i, the relative acceleration of the two particles is given by

$$\frac{d^2\zeta^i}{dt^2} = dg_i = -\Phi_{,i,j}\zeta^j . \tag{8}$$

Thus, the second derivatives of the potential Φ indicate the non-homogeneity of the field. These second derivatives of Φ also satisfy the Poisson's equation, the field equation of Newtonian theory of gravitation

$$\sum_i \Phi_{,i,i} = 4\pi G\rho . \tag{9}$$

At present we shall be interested in vacuum fields — such as the field around the sun (the gravitating mass) — then $\rho = 0$, and Poisson equation reduces to Laplace equation

$$\sum_i \Phi_{,i,i} = 0 . \tag{10}$$

If we compare Eq.(8) with the equation of geodesic deviation as in the previous chapter (equation 68), we see that $R^{\mu}_{\nu\alpha\sigma}u^{\nu}u^{\alpha}$ corresponds to $\Phi_{,i,j}$ and that the following is therefore an analog of Laplace's equation in space-time:

$$R^{\mu}{}_{\nu\rho\mu}u^{\nu}u^{\rho} = 0 \tag{11}$$

summation being implied over the repeated indexes μ, ν, α; and $u^{\nu} = dx^{\nu}/d\tau$. If Eq.(11) is to be held independent of u^{ν}, which after all refers to a specific particle, we need

$$R^{\mu}{}_{\nu\rho\mu} = R_{\nu\rho} = 0 . \tag{12}$$

This is Einstein's law of gravitation in empty space, we shall simply refer it as the vacuum field equations. "Empty" or "vacuum" here means that there is no matter present and no physical fields except the gravitational field. The gravitational field does not disturb the emptiness. Other fields do.

The vacuum field equations (12) lead to

$$R = 0$$

and hence

$$R^{\mu\nu} - \frac{1}{2}g^{\mu\nu}R = 0 \ . \tag{13}$$

If we start with Eq.(13), we get by contraction

$$R - 2R = 0$$

and so we can get back to Eq.(12). We may either use Eq.(12) or Eq.(13) as the basic equations for empty space.

Since the Ricci tensor is symmetric, $R_{\mu\nu} = R_{\nu\mu}$, it has 10 independent components. Accordingly there are 10 vacuum field equations. At first sight the vacuum field equations do not look anything like Newton's. To see a similarity, we must look on the $g_{\mu\nu}$ as a potential describing the gravitational field. Instead of one field equation of Newtonian theory, there are 10 in the Einstein theory. The reason is that the field equations must determine the whole metric, i.e. the $g_{\mu\nu}$, and there are 10 of these (because $g_{\mu\nu} = g_{\nu\nu}$). They describe not only the gravitational field, but also the system of coordinates. The gravitational field and the system of coordinates are inextricably bound up in the Einstein theory, and one cannot describe the one without the other.

The vacuum field equations are like the usual field equations of physics in that they are of the second order, because second derivatives appear in Eq.(12), as the Christoffel symbols involve first derivatives. But the vacuum field equations are unlike the usual field equations in that they are not linear. The nonlinearity means that the equations are complicated and it is difficult to get accurate solutions.

B. Field equations in regions of space where matter is present

The final step in generalizing the Newtonian theory is to write the analog of Poisson's equation (3):

$$\nabla^2\Phi = 4\pi G\rho$$

where ρ is the density of matter, and G the Newton gravitational constant. The left side of Eq.(3) may be made Lorentz-invariant by writing

$$\Box\Phi = 4\pi G\rho \ . \tag{14}$$

Mass, however, is not an invariant, so the right side of the above equation is not a component of four vector. We now recall a conservation law for a fluid in nonrelativistic mechanics, the equation of continuity:

$$\frac{\partial(\rho v^i)}{\partial x^i} + c\frac{\partial\rho}{\partial x^0} = 0 \tag{15}$$

v^i is the velocity, $x^0 = ct$, and the summation over i includes the three space coordinates only. If we multiply Eq.(15) by $dx^1dx^2dx^3$ and integrate over a given volume, the first terms are then transformed to an integral over the closed surface bounding the volume. The result now states that the time derivative of the total mass within the region of integration equals that carried in through the bounding surface.

The special relativistic generalization of Eq.(15) is

$$\frac{\partial T_\mu^{\ \nu}}{\partial x^\nu} = 0 \ . \tag{16}$$

$T_\mu^{\ \nu}$ is the energy-momentum tensor. In Chapter 3 a procedure is available for obtaining $T_\mu^{\ \nu}$ from the Lagrangian density $\mathcal{L}$, and ρc^2 is recognized as the (0,0) component of the energy momentum tensor. Thus, for the right-hand side of the generalized field equations we may consider the energy momentum tensor $T_{\mu\nu}$. The left-hand side must depend purely on the geometric structure, $g_{\mu\nu}$, since we now consider gravitation as a property of space-time alone. Now in the theory of relativity we expect to replace Eq.(16) by

$$T^{\mu\nu}{}_{;\nu} = 0 \tag{17}$$

where the semicolon denotes covariant differentiation. Next, we notice that the covariant divergence of the Ricci tensor $R_{\mu\nu}$ does not vanish, but the covariant divergence of the Einstein tensor

$$G^{\mu\nu} = R^{\mu\nu} - \frac{1}{2}g^{\mu\nu}R \tag{18}$$

does vanish. We therefore write

$$G^{\mu\nu} = \kappa T^{\mu\nu} \tag{19}$$

as the required law of gravitation in regions of space where matter is present. The coupling constant κ is for the present unspecified.

Tracing the Einstein tensor we obtain

$$G_\mu^\mu = g^{\mu\nu}G_{\mu\nu} = R - 2R = -R = \kappa T$$

where $T = T_\mu^\mu$. Therefore the generalized field equations (19) can also be rewritten

$$R_{\mu\nu} = \kappa\left(T_{\mu\nu} - \frac{1}{2}g_{\mu\nu}T\right) . \tag{20}$$

It remains to specify the coupling constant κ in Eqs.(20) and (19). When κ is appropriately chosen, we require that in the limit $c\rightarrow\infty$, Einstein's equations (20) must lead to Poisson's equation. To this aim, let us consider the motion of a particle which follows a geodesic equation

$$\frac{d^2x^\mu}{ds^2} + \Gamma^\mu_{\alpha\beta}\frac{dx^\alpha}{ds}\frac{dx^\beta}{ds} = 0 . \tag{21}$$

In the nonrelativistic limit, neglecting terms of second order in the velocity, we have $ds^2 = c^2dt^2 = (dx^0)^2$, and Eq.(21) reduces to

$$\frac{d^2x^\mu}{dt^2} = -\Gamma^\mu_{00}\left(\frac{dx^0}{dt}\right)^2 = -c^2\Gamma^\mu_{00} . \tag{22}$$

In the limit of weak gravitational field, we can put

$$g_{\mu\nu} = \eta_{\mu\nu} + h_{\mu\nu} \tag{23}$$

with

$$\eta_{00} = 1, \quad \eta_{0i} = 0, \quad \eta_{ij} = \delta_{ij}, \qquad \text{and} \qquad |h_{\mu\nu}| \ll 1 \tag{24}$$

and we can neglect terms of order h^2 and higher. We obtain then for a static field (so $\partial g_{\mu\nu}/\partial x^0 = 0$), from the definition of the Christoffel symbols

$$\Gamma^k_{00} = \frac{1}{2} g^{km}\left(\frac{\partial g_{00}}{\partial x^m}\right) = \frac{1}{2}\frac{\partial g_{00}}{\partial x^k} \tag{25}$$

where the Latin indexes take only three spatial values 1,2,3. The geodesic equation now becomes

$$\frac{d^2\vec{x}}{dt^2} = -\frac{c^2}{2}\nabla h_{00} \tag{26}$$

and coincides with Newton's equation of motion

$$\frac{d^2\vec{x}}{dt^2} = -\nabla\Phi \tag{27}$$

provided that

$$h_{00} = \frac{2}{c^2}\Phi . \tag{28}$$

In the nonrelativistic limit, the leading term in the energy momentum tensor is $T_{00} = \rho c^2$, and the trace is then

$$T = g^{\mu\nu}T_{\mu\nu} \cong g^{00}T_{00} \cong \eta^{00}T_{00} = \rho c^2 . \tag{29}$$

From Eq.(20) we have then

$$R_{00} = \frac{1}{2}\kappa\rho c^2 \,. \tag{30}$$

Neglecting temporal derivatives, and Γ^2 terms, Eq.(4-54) gives in this limit

$$R_{00} = \frac{\partial\Gamma^k_{00}}{\partial x^k} = -\frac{1}{2}\nabla^2 h_{00} = -\frac{1}{c^2}\nabla^2\Phi \,. \tag{31}$$

Combining Eqs.(30) and (31) we finally obtain

$$\nabla^2\Phi = -\frac{1}{2}\kappa\rho c^2 \tag{32}$$

which is identical to the Poisson's equation if we set

$$\kappa = -\frac{8\pi G}{c^4} \,. \tag{33}$$

With this determination of κ, Eq.(19) becomes

$$G^{\mu\nu} \equiv R^{\mu\nu} - \frac{1}{2}g^{\mu\nu}R = -\frac{8\pi G}{c^4}T^{\mu\nu} \tag{34}$$

and the "derivation" of Einstein's field equations (Einstein's laws of gravitation) is completed.

Einstein's field equations (34) need not be supplemented by any statement concerning equations of motion; these are given by

$$T^{\mu\nu}{}_{;\nu} = 0 \tag{35}$$

How is the energy momentum tensor to be specified in general relativity for a given physical system? The prescription for this is the following: The required expression for $T^{\mu\nu}$ is the covariant generalization of the one valid in the special relativity. For example, for a perfect fluid described in terms of an energy density ϵ and a scalar pressure p, the correct expression for its energy momentum tensor is

$$T^{\mu\nu} = (\epsilon + p)u^\mu u^\nu - g^{\mu\nu}p, \qquad \left(u^\nu = \frac{dx^\nu}{ds}\right) \tag{36}$$

because this expression is the correct one in special relativity. The justification for the above prescription is that since we can always set up a Galilean coordinate frame locally and since the Einstein tensor is invariant to general coordinate transformations (i.e., covariant) the expression for $T^{\mu\nu}$ in a frame that is not locally Galilean should be obtainable by subjecting $T^{\mu\nu}$ (valid in the local Galilean frame) to the same tensor transformations.

One last point should be mentioned. In a curved space the conservation of energy and momentum is only approximate. The reason is attributed to the gravitational field which itself works on the matter and has itself some energy and momentum.

5-3 Equations of motion

It was stated in the preceding section that Einstein's field equations need not be supplemented by equations of motion; these are given by Eq.(35)

$$T^{\mu\nu}{}_{;\nu} = 0 \, .$$

Since the Einstein tensor $G^{\mu\nu}$ (= $R^{\mu\nu}$ - $(1/2)g^{\mu\nu}R$) has zero covariant divergence, Einstein's field equations necessarily require that the covariant divergence of $T^{\mu\nu}$ vanish. Thus, Einstein's field equations contain the equations of motion. This is in contrast to the situation in electrodynamics, where Maxwell's equations lack the corresponding equation of motion. It remains to be demonstrated that Eq.(35) is indeed an equation of motion. To this end, let us consider the energy momentum tensor of a single particle of rest mass m

$$T^{\mu\nu}(x) = \left(m/\sqrt{-g}\right)\int u^{\mu}u^{\nu}\delta^4(x - y(s))ds \; ; \qquad u^{\mu} = \frac{dy^{\mu}}{ds} \tag{37}$$

where the integral is taken over the particle's trajectory; and

$$\delta^4(x) = \delta(x^0)\delta(x^1)\delta(x^2)\delta(x^3)$$

$\delta(x)$ is the δ-function defined by the relations

$$\delta(x) = \infty \quad \text{if} \quad x = 0; \qquad \text{and} \qquad \delta(x) = 0 \quad \text{if} \quad x \neq 0$$

so that

$$\int_a^b \delta(x) = 1, \qquad \text{where} \qquad a < 0 < b \tag{38}$$

The basic property of the δ-function

$$\int_a^b f(x)\delta(x - x_0)\,dx = f(x_0), \qquad a < x_0 < b \tag{39}$$

follows immediately from the above definition, where $f(x)$ is an arbitrary continuous function of x.

Note that in writing Eq.(37) use has been made of the fact that $\delta^4(x - y)/\sqrt{-g}$ is a general scalar and that, in the absence of gravity, Eq.(37) reduces to the familiar expression in special relativity. Before we insert Eq.(37) into Eq.(35), let us first carry out the covariant differentiation of $T^{\mu\nu}$:

$$T^{\mu\nu}{}_{;\nu} = \frac{1}{\sqrt{-g}}\frac{\partial}{\partial x\lambda}\left(\sqrt{-g}\,T^{\mu\nu}\right) + \Gamma^{\mu}_{\nu\lambda}T^{\nu\lambda} . \tag{40}$$

Now inserting Eq.(37) into Eq.(40), we obtain

$$\int u^{\mu}u^{\nu}\frac{\partial}{\partial x^{\nu}}\delta^4(x - y(s))ds + \Gamma^{\mu}_{\lambda\nu}\int u^{\lambda}u^{\nu}\delta^4(x - y(s))ds = 0 . \tag{41}$$

To evaluate the first term in the above, note that

$$\begin{aligned} &\int u^{\mu}u^{0}\delta^3(x - y)\frac{\partial}{\partial x^{0}}\delta(x^{0} - y^{0})ds \\ &= \int u^{\mu}\delta^3(x - y)\frac{\partial}{\partial x^{0}}\delta(x^{0} - y^{0})dy^{0} \\ &= \frac{d}{dx^{0}}(u^{\mu}\delta^3(x - y)) . \end{aligned} \tag{42}$$

Similarly

$$\int u^{\mu} u^{i} \delta(x^{0} - y^{0}) \frac{\partial}{\partial x^{i}} \delta^{3}(x - y) ds$$

$$= -\int u^{\mu} \delta(x^{0} - y^{0}) \frac{\partial}{\partial y^{i}} \delta^{3}(x - y) dy^{i}$$

$$= -\int u^{\mu} \delta(x^{0} - y^{0}) \frac{dy^{i}}{dy^{0}} \frac{\partial}{\partial y^{i}} \delta^{3}(x - y) dy^{0} \qquad (43)$$

$$= -\int u^{\mu} \delta(x^{0} - y^{0}) \frac{\partial}{\partial y^{0}} \delta^{3}(x - y) dy^{0}.$$

$$= -u^{\mu} \frac{\partial}{\partial x^{0}} \delta^{3}(x - y(x^{0})).$$

Adding Eqs.(42) and (43) we obtain

$$\int u^{\mu} u^{\nu} \frac{\partial}{\partial x^{\nu}} \delta^{4}(x - y(s)) ds = \frac{du^{\mu}}{ds} \delta^{3}(x - y)$$

$$= \int du^{\mu} \delta^{4}(x - y) = \int \frac{du^{\mu}}{ds} \delta^{4}(x - y(s)) ds \ . \qquad (44)$$

Inserting Eq.(44) into Eq.(41), we obtain

$$\int \left(\frac{du^{\mu}}{ds} + \Gamma^{\mu}_{\lambda\nu} u^{\lambda} u^{\nu} \right) \delta^{4}(x - y(s)) ds = 0 \qquad (45)$$

from which it follows the equations of motion

$$\frac{du^{\mu}}{ds} + \Gamma^{\mu}_{\lambda\nu} u^{\lambda} u^{\nu} = 0 \ . \qquad (46)$$

This analysis discussion sheds some light on the equations of motion. You might have noticed that the geodesic postulate and Einstein's field equations are introduced independently of each other and there seems to be no logical necessity that one should depend on the other. The above discussion clearly demonstrates that, on the contrary, they are not quite independent. We may regard the right-hand side, $T^{\mu\nu}$, of Einstein's field equations as the result of the computation of the purely "geometrical" tensor of the left-hand side, the Einstein tensor. This would amount to a redefinition of the energy-momentum tensor, $T^{\mu\nu}$, solely in terms of geometry. In particular, the point concentration of energy (particles) would now be regarded as a singularity of the geometrical field $g_{\mu\nu}$. Then it follows that Einstein's field equation, now a purely geometrical equation, should determine how these energy concentrations move. If the theory of general relativity is totally consistent, the equations of motion for these energy concentrations must turn out to be identical with the geodesic equations of motion. The rigorous proof of this consistency requirement turns out to be quite difficult, especially if the energy concentrations are not exactly point-like. In 1938, Einstein, Infeld, and Hoffmann carried out a series of successive approximations and proved, to third order in M/r, that the geodesic equations follow from the field equations. Since then, the calculations have been carried to sixth order by Infeld and his associates, who have found that the geodesic postulates still hold good. It is now generally believed that for small energy concentrations Einstein's field equations necessarily lead to geodesic equations of motion without any further assumptions.

The equation of motion (46) can be put into a different form which is very useful in determining the constants of motion. Multiplying Eq.(46) by $g_{\mu\nu}$, putting in the expression (4-42) for $\Gamma^{\mu}_{\lambda\nu}$ and using Eq.(4-13), we obtain

$$g_{\mu\nu}\frac{du^{\nu}}{ds} + \left(\frac{\partial g_{\mu\nu}}{\partial x^{\lambda}} - \frac{1}{2}\frac{\partial g_{\nu\lambda}}{\partial x^{\mu}}\right)u^{\nu}u^{\lambda} = 0 \tag{47}$$

but

$$\frac{du_{\mu}}{ds} = \frac{d}{ds}\left(g_{\mu\nu}u^{\nu}\right) = g_{\mu\nu}\frac{du^{\nu}}{ds} + \frac{\partial g_{\mu\nu}}{\partial x^{\lambda}}u^{\lambda}u^{\nu} = 0$$

Substituting this into Eq.(47), we obtain

$$\frac{du_\mu}{ds} - \frac{1}{2}\frac{\partial g_{\nu\lambda}}{\partial x^\mu}u^\nu u^\lambda = 0 \tag{48}$$

which shows clearly that if $\partial g_{\nu\lambda}/\partial x^\mu = 0$ for all ν and λ, $du_\mu/ds = 0$. That is, if the components of the metric tensor do not depend on a certain coordinate x^μ, then the corresponding covariant component of velocity u_μ is a constant of motion. These conserved quantities are often called the first integrals of the equation of motion (46).

Problems

(5-1) Prove the result shown in Eq.(31).

(5-2) Given the space-time geometry

$$ds^2 = e^{-2\Phi}dt^2 - e^{2\Phi}(dx^2 + dy^2 + dz^2)$$

find the geodesic equations of motion.

(5-3) If space-time has the metric

$$ds^2 = e^{\lambda}(dr^2 + dz^2) + r^2e^{-\rho}d\Phi^2 - e^{\rho}dt^2$$

where λ, ρ are functions of r and z only, show that the field equations in empty space R = 0 require that the following are satisfied

$$\lambda_1 + \rho_1 = \frac{1}{2}r\left(\rho_1^2 - \rho_2^2\right)$$

$$\lambda_2 + \rho_2 = r\rho_1\rho_2$$

$$\rho_{11} + \rho_{22} + \frac{1}{r}\rho_1 = 0$$

$$\lambda_{11} + \lambda_{22} + \rho_{11} + \rho_{22} + \frac{1}{2}\left(\rho_1^2 + \rho_2^2\right) = 0$$

where subscripts 1 and 2 denote partial differentiations with respect to r and z respectively.

(5-4) If space-time has the metric

$$ds^2 = e^{2kx}(dx^2 + dy^2 + dz^2 - dt^2)$$

where k is constant, and $v^2 = x^2 + y^2 + z^2$, dots denoting differentiations with respect to t, show that for a freely falling body

$$1 - v^2 = (1 - V^2)e^{2kx}$$

where v = V at x = 0.

Problems

Chapter 6

The Schwarzschild Solution

Einstein's field equations are nonlinear and are therefore very complicated, and it is difficult to get accurate solutions of them. There is, however, one special case which can be solved without too much trouble; namely, the static spherically symmetric field produced by a spherical mass M at rest. This was first found in 1916 by Schwarzschild, and this solution has played a major role in the early development of general relativity and is even today regarded as a solution of fundamental importance. In the Newtonian gravitation the solution of this problem is described by a gravitational potential $\Phi = GM/r$, where M is the gravitational mass of the source distribution and r is the radius coordinate from the center of the mass distribution. The Schwarzschild solution describes the general relativistic analogue of the Newtonian solution.

6-1 The Schwarzschild metric

The static condition means that, with a static coordinate system, the g_{ij} are independent of the time x^0 or t and also $g_{0i} = 0$. The spatial coordinates may be taken to be spherical polar coordinates $x^1 = r$, $x^2 = 0$, $x^3 = 0$. The most general form for ds^2 compatible with spherical symmetry is

$$ds^2 = A\,c^2dt^2 + 2Hdtdr + B\,dr^2 + C\,r^2(d\theta^2 + \sin^2\theta\, d\varphi^2), \qquad (1)$$

where A, H, B, and C are general functions of r and t. The subspace r = constant, t = constant are spheres on which a typical point is labelled by the coordinates θ and φ. The center of symmetry is the point 0 given by r = 0. We may replace r by any function of r without disturbing the spherical symmetry. We now use this freedom to simplify things as much as possible. Consider a coordinate transformation:

$$r = r', \qquad t = k(r',t') \tag{2}$$

where k is to be chosen as follows. From this coordinate transformation we have

$$dr = dr', \qquad dt = \frac{\partial k}{\partial r'}dr' + \frac{\partial k}{\partial t'}dt' \tag{3}$$

and accordingly ds^2 becomes

$$ds^2 = A\left(\frac{\partial k}{\partial r'}dr' + \frac{\partial k}{\partial t'}dt'\right)^2 + 2Hdr'\left(\frac{\partial k}{\partial r'}dr' + \frac{\partial k}{\partial t'}dt'\right) + Bdr'^2 + C(d\theta^2 + \sin^2\theta d\varphi^2) \tag{4}$$

In this the coefficient of $dr'dt'$ is zero provided

$$\left(A\frac{\partial k}{\partial r'} + H\right)\frac{\partial k}{\partial t'} = 0 \tag{5}$$

In general $\partial k/\partial t' = 0$; otherwise there is no inverse of the transformation (2). Hence if we choose k to satisfy the relation

$$A\frac{\partial k}{\partial r'} + H = 0 \tag{6}$$

we can rewrite the line-element ds^2 (1) in the form

$$ds^2 = e^{2\sigma}dt^2 - e^{2\omega}dr^2 - e^{2\mu}(d\theta^2 + \sin^2\theta\, d\varphi^2) \tag{7}$$

where σ, ω and μ are real functions of r′, t′. We have written the metric components in exponential form to ensure the right signature. So Eq.(7) may be considered as the most general spherically symmetric line-element.

We now introduce a new radial coordinate R, the Schwarzschild radial coordinate:

$$R = e^{\mu/2}. \qquad (8)$$

To preserve the diagonality of the metric tensor, we need to introduce a new Schwarzschild time coordinate T by a transformation of the type

$$t' = h(R,T) \qquad (9)$$

where the h is chosen to make the coefficient of dRdT in ds^2 zero. The expression for ds^2 may then be written in the form

$$ds^2 = e^{2\nu}dt^2 - e^{2\lambda}dr^2 - r^2(d\theta^2 + \sin^2\theta d\varphi^2) \qquad (10)$$

where, instead of R and T, r and t are the Schwarzschild radial and time coordinates, respectively. The functions μ and λ depend on r and t in general. The Schwarzschild radial coordinate has a special meaning: the area of the surface of the sphere, r = constant, is given by $4\pi^2r^2$. In the future we will refer to r as a Schwarzschild radial coordinate whenever we want to imply the above result.

6-2 *The Schwarzschild solution of the vacuum field equations*

We now work out the Einstein field equations in detail. We can immediately read off the values of the $g_{\mu\nu}$ from Eq.(10):

$$g_{00} = e^{2\nu}, \quad g_{11} = -e^{2\lambda}, \quad g_{22} = -r^2, \quad g_{33} = -r^2\sin^2\theta, \qquad (11)$$

$$g_{\mu\nu} = 0, \quad \text{for} \quad \mu \neq \nu \qquad (12)$$

and we find

$$g^{00} = e^{-2\nu}, \quad g^{11} = -e^{-2\lambda}, \quad g^{22} = -r^{-2}, \quad g^{33} = -r^{-2}\sin^{-2}\theta, \qquad (13)$$

$$g^{\mu\nu} = 0, \quad \text{for} \quad \mu \neq \nu\ . \qquad (14)$$

With these values it is easy to calculate the $\Gamma^{\alpha}_{\beta\gamma}$ from the following formula

$$\Gamma^{\alpha}_{\beta\gamma} = \frac{1}{2} g^{\alpha\delta}\left(\frac{\partial g_{\delta\beta}}{\partial x^{\gamma}} + \frac{\partial g_{\delta\gamma}}{\partial x^{\beta}} - \frac{\partial g_{\beta\gamma}}{\partial x^{\delta}} \right).$$

The calculation leads to the following expressions, with primes denoting differentiations with respect to r,

$$\Gamma^{1}_{00} = \nu' e^{2\nu - 2\lambda}, \quad \Gamma^{1}_{11} = \lambda', \quad \Gamma^{1}_{22} = -re^{-2\lambda}, \quad \Gamma^{1}_{33} = -r\sin^2\theta e^{-2\lambda} \tag{15}$$

and

$$\Gamma^{0}_{10} = \nu', \quad \Gamma^{2}_{12} = \Gamma^{3}_{13} = r^{-1}, \quad \Gamma^{3}_{23} = \cot\theta, \quad \Gamma^{2}_{33} = -\sin\theta\cos\theta \tag{16}$$

All other components (except for those which differ from the ones we have written by a transposition of the indices β and γ) are zero. These expressions are to be substituted in the following formula:

$$R_{\mu\nu} = \frac{\partial \Gamma^{\alpha}_{\mu\alpha}}{\partial x^{\nu}} - \frac{\partial \Gamma^{\alpha}_{\mu\nu}}{\partial x^{\alpha}} - \Gamma^{\alpha}_{\mu\nu}\Gamma^{\beta}_{\alpha\beta} + \Gamma^{\alpha}_{\mu\beta}\Gamma^{\beta}_{\nu\alpha}$$

and the results are

$$R_{00} = \left(-\nu'' + \lambda'\nu' - \nu'^2 - \frac{2\nu'}{r} \right) e^{(2\nu - 2\lambda)}, \tag{17}$$

$$R_{11} = \nu'' - \lambda'\nu' + \nu'^2 - \frac{2\lambda'}{r} \tag{18}$$

$$R_{22} = (1 + r\nu' - r\lambda')e^{-2\lambda} - 1, \tag{19}$$

$$R_{33} = R_{22} \sin^2\theta, \tag{20}$$

with the other components of $R_{\mu\nu}$ vanishing.

Einstein's law of gravity in empty space requires these expressions to vanish. The vanishing of Eqs.(17) and (18) leads to

$$\lambda' + \nu' = 0 .$$

For large values of r the space must approximate to being flat, so that λ and ν both tend to zero as $r \longrightarrow \infty$. It follows that

$$\lambda + \nu = 0 . \tag{21}$$

The vanishing of Eq.(19) now gives

$$(1 + 2r\nu')e^{2\nu} = 1$$

or

$$(re^{2\nu})' = 1$$

$$re^{2\nu} = r - 2m \tag{22}$$

where m is an integration constant. This also makes Eqs.(17) and (18) vanish. We now get

$$g_{00} = 1 - 2m/r . \tag{23}$$

The constant m can be expressed in terms of the mass of the body by requiring that Newton's law should hold at large distances where the field is weak. In other words, we should have $g_{00} = 1 + 2\Phi/c^2$, where the potential has its Newtonian value $\Phi = -GM/r$. From this it is clear that $m = -2GM/c^2$. This quantity has the dimension of length; it is called the gravitational radius r_g of the body: $r_g = 2GM/c^2$. Thus

$$e^{2\nu} = g_{00} = 1 - 2GM/rc^2 = 1 - r_g/r \tag{24}$$

and

$$e^{2\lambda} = e^{-2\mu} = g_{11} = (1 - r_g/r)^{-1} . \tag{25}$$

Thus we finally obtain the spacetime metric in the form

$$ds^2 = (1 - r_g/r)c^2dt^2 - (1 - r_g/r)^{-1}dr^2 - r^2(d\theta^2 + \sin^2\theta\, d\varphi^2. \tag{26}$$

This solution of the Einstein equations was found by Schwarzschild (1916). It tells us the important result: "the empty spacetime outside a spherically symmetric distribution of matter is describable by a static metric." This result is known as **Birkhoff's theorem**. We emphasize that the Schwarzschild solution is also valid for masses which are moving, so long as the motion has the required symmetry, for example a centrally-symmetric pulsation. We note that the metric (26) depends only on the total mass of the gravitating body, just as in the analogous problem in Newtonian theory.

If we set dt = 0 in the Schwarzschild metric, we obtain a 3-dimensional manifold with line element

$$-dL^2 = \left(1 - \frac{r_g}{r}\right)^{-1} dr^2 + r^2d\theta^2 + r^2\sin^2\theta d\varphi^2 \tag{27}$$

which can be put into the form

$$dL^2 = g_{ij}dx^idx^j, \quad (i,j = 1,2,3;\ x^1 = r,\ x^2 = \theta,\ x^3 = \varphi) \tag{28}$$

we see that no g_{ij} depends on t, so the slice given by t = constant is a space rather than a spacetime. This space allows us to refer to events with the same r, θ, ϕ-coordinates, but different time coordinates, as occurring at the same point in space. This splitting of spacetime into space and time is possible in any <u>static</u> spacetime, <u>but is not a feature of spacetimes in general</u>.

If we can turn M (which appears in $r_g = 2GM/c^2$) down to zero, then the line element (26) becomes that of flat spacetime in spherical polar coordinates, and the line element (27) becomes that of Euclidean space in spherical polar coordinates. Turning up M introduces distortion into both spacetime and space, so that neither is flat. This distortion is measured by the dimensionless quantity r_g/r. In the flat spacetime, the coordinate r is the distance from the origin, but r no longer measures radial distance in a curved spacetime. What then does r represent? If we take the sphere in space given by r = constant, then we have, by setting dr = 0 in the line element

$$dL^2 = r^2(d\theta^2 + \sin^2\theta d\varphi^2) \tag{29}$$

which shows that this sphere has the 2-dimensional geometry of a sphere of radius r embedded in euclidean space, and just as in the flat space, infinitesimal tangential distances are given by

$$dL = r(d\theta^2 + \sin^2\theta d\varphi^2)^{1/2} . \tag{30}$$

The line element (27) shows that the infinitesimal radial distance given by θ and φ constant is

$$dR = (1-r_g/r)^{-1/2}dr \tag{31}$$

so $dR > dr$. The distance between two points r_1 and r_2 along the same radius is given by the integral

$$\int_{r_1}^{r_2} \frac{dr}{\sqrt{1 - r_g/r}} > r_2 - r_1 . \tag{32}$$

The quantity r is often called the "angular diameter distance". The apparent incompatibility of the distances (30) and (31) is explained by the curvature of space. Figure 1 shows radial distance dR in the curved space, where the flat disc S_0 represents a portion of the flat space and S_m represents a portion of the curved space. The circles C_1 and C_3 represent spheres having the geometry of a sphere of radius r in Euclidean space, and C_2 and C_4 represent neighboring spheres having the geometry of a sphere of radius r + dr in Euclidean space. We can see that, in the flat space represented by S_0, the measured radial distance between the spheres is dr; but in the curved space represented by S_m the measured distance is dR given by Eq.(31). If we were to measure the circumference of a great circle of the sphere r = constant using small measuring rods, the same number of rods would be needed in flat space as in the curved space. But if we were to measure the radial distance between points with radial coordinates r_1 and r_2, then more rods would be needed in the curved space than in the flat space (Figure 2).

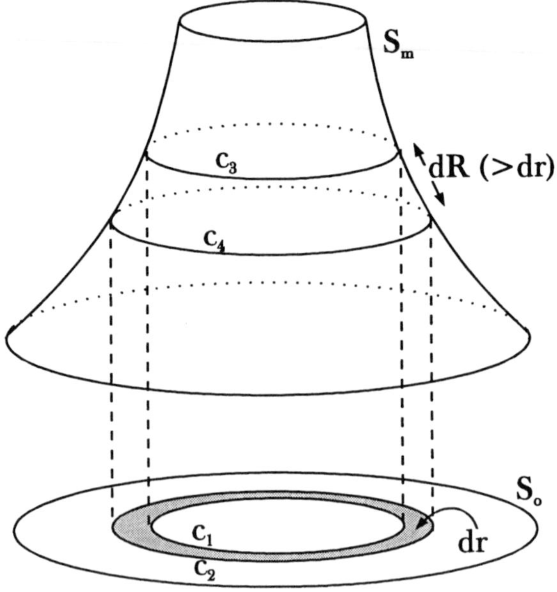

Figure 1: Radial distance in the Schwarzschild geometry

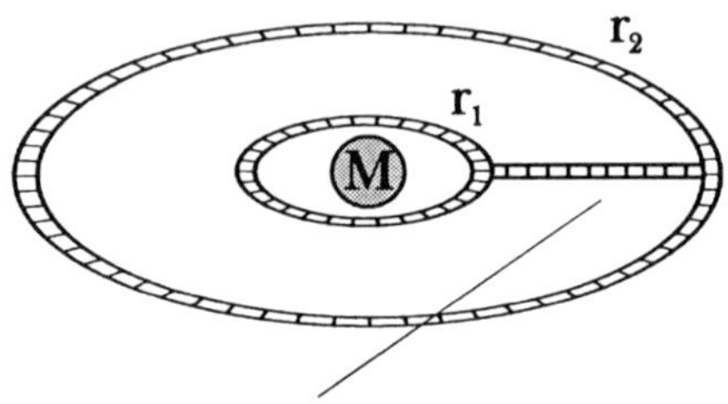

Figure 2: Measuring distance in the Schwarzschild geometry

We now turn our attention to the measurement of times in the curved spacetime. One of the basic assumptions taken from special relativity is that clocks record proper time intervals along their world lines. For a clock at a fixed point in space (r, θ, φ constant) the infinitesimal proper time interval is given by

$$d\tau = g_{00}dt = \sqrt{1 - r_g/r}\, dt \tag{33}$$

so in flat spacetime $d\tau = dt$, and a fixed clock records the coordinate time t. But in the curved spacetime, fixed clocks record only proper time, and because $g_{00} < 1$,

$$d\tau \leq dt \ . \tag{34}$$

The equality sign holds only at infinity, where t coincides with the proper time. Thus at finite distances from the masses there is a "slowing down" of the time compared with the time at infinity. If a light of certain frequency is emitted and is being observed at a large value of r, the light is red-shifted by a factor $g_{00}^{-1/2}$.

One final point to note is that as $r \longrightarrow \infty$, $dR \longrightarrow dr$ and $d\tau \longrightarrow dt$, so asymptotically the coordinate distance dr coincides with the actual distance dR, the coordinate time dt with the proper time $d\tau$, and the line element dS^2 becomes

$$ds^2 = ds_0^2 - (2GM/c^2r)(dr^2 + c^2dt^2) \ . \tag{35}$$

The second term represents a small correction to the Galilean metric ds_0^2. At large distances from the masses producing it, every field appears centrally symmetric. Therefore, Eq.(35) determines the metric at large distances from any system of bodies.

The solution (26) becomes singular at $r = r_g$, because $g_{00} = 0$ and g_{11} approaches $-\infty$. For ordinary bodies the gravitational radius r_g (also known as the Schwarzschild radius) lies well inside them, where the Schwarzschild solution is inapplicable anyway; for example, for our sun it is 2.9 km, for the earth, 0.88 cm, for a proton, 2.4×10^{-52} cm. We will see in Chapter 8 that the singularity which appears at $r = r_g$ is more a consequence of the choice of the coordinate frame than of the geometry itself. Nevertheless, interesting things do happen at the boundary $r = r_g$. For example, matter and energy may fall into the region $r < r_g$ but nothing (including light signal) can come out of it. Such a region is called a **black hole**, and we will discuss this situation in Chapter 8.

Problems

(6-1) Prove the results of Eqs.(15) and (16).

(6-2) Prove the results of Eqs.(17) to (20).

(6-3) Calculate the gravitational radius (or Schwarzschild radius) for (a) a galaxy ($M \simeq 10^{11} M_{\odot}$), (b) a proton.

(6-4) Show that the transformation

$$r = r'\left(1 + \frac{GM}{2c^2 r'}\right)^2$$

puts the Schwarzschild metric (26) into the **isotropic** form

$$ds^2 = \left(1 + \frac{GM}{2c^2 r'}\right)[dr'^2 + r'^2(d\theta^2 + \sin^2\theta\, d\varphi^2)] - \left(\frac{1 - (GM/2c^2 r')}{1 + (GM/2c^2 r')}\right)^2 c^2 dt^2 \ .$$

(6-5) Resolve the clock paradox of special relativity by using the above isotropic metric.

(6-6) Find the equations determining the static gravitational field in a vacuum around an axially symmetric body at rest.

Chapter 7

Experimental Tests of Einstein's Theory

The Schwarzschild metric enables the behavior of test bodies, light rays and clocks to be predicted, so that Einstein's theory of gravitation can be tested. We shall describe some of these tests in this chapter.

7-1 Precession of the perihelion of Mercury

According to Kepler's Second Law the orbit of a planet is a closed ellipse with the sun at one of the two foci, and so its perihelion (the point closest to the sun) is fixed. The relativistic Newtonian mechanics predicts that the perihelion is not fixed, but gradually rotates around the sun. However, it only gives 1/6 of the observed centennial precession.

We now consider this problem in the framework of Einstein's theory. Consider the motion of Mercury in the space-time continuum characterized by the Schwarzschild metric. We assume that the mass of Mercury is negligible in comparison with the mass of the sun ($M_\odot = 2 \times 10^{30}$ kg, $M_{☿} = 3.3 \times 10^{23}$ kg), so that its presence does not modify the metric to any appreciable extent. Then the motion of Mercury is a time-like geodesic in the Schwarzschild space-time surrounding the sun:

$$\frac{d^2x^\mu}{ds^2} + \Gamma^\mu_{\nu\lambda}\frac{dx^\nu}{ds}\frac{dx^\lambda}{ds} = 0 \tag{1}$$

Taking $x^1 = r$, $x^2 = \theta$, and $x^3 = \varphi$, Eq.(1) then becomes

$$\frac{d^2r}{ds^2} + \frac{1}{2}\frac{d\lambda}{dr}\left(\frac{dr}{ds}\right)^2 - re^{-2\lambda}\left(\frac{d\theta}{ds}\right)^2 - r\sin^2\theta e^{-2\lambda}\left(\frac{d\varphi}{ds}\right)^2 + \frac{1}{2}e^{2(\nu-\lambda)}\frac{d\nu}{dr}\left(c\frac{dt}{ds}\right)^2 = 0 \quad (2)$$

$$\frac{d^2\theta}{ds^2} + \frac{2}{r}\frac{dr}{ds}\frac{d\theta}{ds} - \sin\theta\cos\theta\left(\frac{d\varphi}{ds}\right)^2 = 0 \quad (3)$$

$$\frac{d^2\varphi}{ds^2} + \frac{2}{r}\frac{dr}{ds}\frac{d\varphi}{ds} + 2\cot\theta\frac{d\theta}{ds}\frac{d\varphi}{ds} = 0 \,. \quad (4)$$

and

$$\frac{d^2t}{ds^2} + 2\frac{d\nu}{dr}\frac{dr}{ds}\frac{dt}{ds} = 0 \,. \quad (5)$$

Due to the spherical spatial symmetry of the Schwarzschild metric, conservation of angular momentum still holds and the motion of the planet remains in a plane. The plane in which the initial velocity vector and the initial position vector of the planet lie may be taken as the equatorial plane ($\theta = \pi/2$). Of course, initially $d\theta/ds = 0$; then by Eq.(3), $d^2\theta/ds^2$ is also zero. Consequently, θ remains constant throughout the motion, and the orbit is permanently confined to the equatorial plane. We may simplify all the equations, putting everywhere $\theta = \pi/2 =$ constant, and also remembering that $\lambda = -\nu$:

$$\frac{d^2r}{ds^2} - \frac{1}{2}\frac{d\nu}{dr}\left(\frac{dr}{ds}\right)^2 - re^{2\nu}\left(\frac{d\varphi}{ds}\right)^2 + \frac{1}{2}e^{4\nu}\frac{d\nu}{dr}\left(c\frac{dt}{ds}\right)^2 = 0 \quad (6)$$

$$\frac{d^2\varphi}{ds^2} + \frac{2}{r}\frac{dr}{ds}\frac{d\varphi}{ds} = 0 \quad (7)$$

and

$$\frac{d^2t}{ds^2} + 2\frac{d\nu}{dr}\frac{dr}{ds}\frac{dt}{ds} = 0 \tag{8}$$

where

$$e^{2\nu} = 1 - \frac{2GM}{c^2 r} . \tag{9}$$

Now, Eq.(7) can be written as

$$\frac{1}{r^2}\frac{d}{ds}\left(r^2\frac{d\varphi}{ds}\right) = 0$$

which is readily integrated and gives

$$r^2\frac{d\varphi}{ds} = \text{constant} = h . \tag{10}$$

This result is the relativistic analogue of Kepler's Second Law of planetary motion, the constant h being a measure of the angular momentum of the motion concerned.

Similarly, Eq.(8) can be written as

$$\frac{d}{ds}\left(e^{2\nu}\frac{dt}{ds}\right) = 0$$

which is readily integrated and gives

$$\frac{dt}{ds} = K e^{-2\nu} \tag{11}$$

where K is an integration constant. For the r-equation we may use, instead of Eq.(6), a simpler equation directly obtainable from the Schwarzschild metric (with $\theta = \pi/2$ and $d\theta = 0$):

$$1 = \gamma\left(\frac{dt}{ds}\right)^2 - \gamma^{-1}\left(\frac{dr}{ds}\right)^2 - r^2\left(\frac{d\varphi}{ds}\right)^2 \tag{12}$$

where we have put $r = e^{2\nu} = 1 - 2GM/c^2r$. Using Eqs.(10) and (11) this becomes

$$1 = \frac{K^2}{\gamma} - \frac{h^2}{r^2} - \frac{1}{\gamma}\left(\frac{h}{r^2}\frac{dr}{d\varphi}\right)^2 . \tag{13}$$

Multiplying by γ and putting $W = r^{-1}$, we find

$$\left(\frac{dW}{d\varphi}\right)^2 + W^2 = \frac{K^2 - 1}{h^2} + \frac{2m}{h^2}W + 2mW^3 \tag{14}$$

where we have put $m = 2GM/c^2$. Differentiating Eq.(14) with respect to φ, we obtain two solutions: one is $dW/d\varphi = 0$, corresponding to $r =$ constant, i.e. to a circular orbit. In the other case $dW/d\varphi \neq 0$, corresponding to an orbit described by the following differential equation

$$\frac{d^2W}{d\varphi^2} + W = \frac{GM}{c^2h^2} + \frac{3GM}{c^2}W^2 \tag{15}$$

which differs from its Newtonian counterpart by the presence of the nonlinear term $(3GM/c^2)W^2$. This nonlinear term vanishes as $c \to \infty$, as expected.

We now take a brief digression to obtain the Newtonian equation of the orbit. We can start with the two-body Lagrangian which is given by

$$L = \frac{1}{2}\left(\dot{r}^2 + r^2\dot{\varphi}^2\right) + \frac{GM}{r} \tag{16}$$

where a dot denotes differentiation with respect to t. The Lagrange equation for φ then gives

$$\frac{d}{dt}\frac{\partial L}{\partial \dot{\varphi}} = \frac{d}{dt}\left(r^2\dot{\varphi}\right) = 0$$

from which we obtain

$$r^2\dot{\varphi} = hc \tag{17}$$

where h is an integration constant, and c is the speed of light. As the

Lagrangian is not explicitly time-dependent, the total energy E is also a constant of the motion:

$$E = \frac{1}{2}\left(\dot{r}^2 + r^2\dot{\varphi}^2\right) - \frac{GM}{r}$$

$$= \frac{1}{2}\left(\dot{r}^2 + \frac{h^2c^2}{r^2}\right) - \frac{GM}{r} \qquad (18)$$

Putting $\dot{r} = (dr/d\varphi)\dot{\varphi}$ and $W = r^{-1}$, differentiating with respect to φ and assuming $dW/d\varphi \neq 0$, we obtain, from Eq.(18), the following equation of the orbit

$$\frac{d^2W}{d\varphi^2} + W = \frac{GM}{c^2h^2}\ . \qquad (19)$$

It is easy to see that this equation differs from the general relativistic equation of the orbit (15) by the nonlinear term. The general relativistic correction is very small, and for ordinary speeds it induces a negligible deviation from the Newtonian orbit. In fact, the ratio between $3GMW^3/c^2$ and GMW/c^2 is $3h^2W^2$, that is $3(r\dot{\varphi}/c)^2$; it thus represents three times the square of the transverse velocity of the planet measured in units of c, and for earth, for example, the orbital speed is ~30 km/sec, the ratio is $\sim 10^{-8}$. However, because of this small additional term, the planetary orbits are no longer closed, and a precession of the perihelion results. To see this effect, let us solve the general relativistic equation of orbit (14) in successive orders by an approximation procedure. We take the solution of the Newtonian equation of the orbit as the zero-order unperturbed solution which is

$$W^{(0)} = \frac{m}{h^2}\left[1 + e\cos(\varphi - \varphi_0)\right] \qquad (20)$$

where $m = GM/c^2$, e and φ_0 are integration constants. For $0 < e < 1$, this solution represents an ellipse of eccentricity e. We can orient the coordinate system so that $\varphi_o = 0$. The length of the major axis of the ellipse, denoted by 2a, is given by

$$2a = \left(\frac{1}{W^{(0)}}\right)_{\varphi = 0} + \left(\frac{1}{W^{(0)}}\right)_{\varphi = \pi} . \tag{21}$$

Using this relation, we can write h^2 in terms of the eccentricity e and semimajor axis a:

$$h^2 = \frac{aGM}{c^2}(1 - e^2) . \tag{22}$$

In the first approximation, Eq.(15) becomes

$$\frac{d^2W^{(1)}}{d\varphi^2} + W^{(1)} = \frac{m}{h^2} + 3m(W^{(0)})^2 . \tag{23}$$

Substituting Eq.(20) into this equation, we obtain

$$\frac{d^2W^{(1)}}{d\varphi^2} + W^{(1)} = \frac{3m^3}{h^4} + \frac{6m^3}{h^4} e \cos(\varphi - \varphi_0) + \frac{3m^3}{h^4} e^2 \cos^2(\varphi - \varphi_0) . \tag{24}$$

Considering orbits with small eccentricity (like the planet Mercury), the terms of order e^2 and higher may be neglected, in the first approximation. The contribution of the constant term $3m^3/h^4$ is also negligible. The one that produces an observable effect is the term in $\cos(\varphi - \varphi_0)$, whose contribution increases continuously at each revolution. Therefore we now have

$$\frac{dW^{(1)}}{d\varphi^2} + W^{(1)} = \frac{6m^3}{h^4} e \cos(\varphi - \varphi_0) \tag{25}$$

and the solution is

$$W^{(1)} = \frac{3m^3}{h^4} e\, \varphi \sin(\varphi - \varphi_0) . \tag{26}$$

Including this correction, the solution of the general relativistic orbital equation can now be written as

$$W \simeq W^{(0)} + W^{(1)} = \frac{m}{h^2}\left[1 + e\cos(\varphi - \varphi_0) + \frac{3m^2}{h^2} e\,\varphi\sin(\varphi - \varphi_0)\right] . \quad (27)$$

Now let

$$\Delta\varphi_o = \frac{3m^2}{h^2}\varphi \quad (28)$$

and note that

$$\cos(\varphi - \varphi_0 - \Delta\varphi_0) = \cos(\varphi - \varphi_0)\cos\Delta\varphi_0 + \sin(\varphi - \varphi_0)\sin\Delta\varphi_0 \quad (29)$$

Since $\Delta\varphi_0 << 1$,

$$\cos(\Delta\varphi_0) \simeq 1, \qquad \sin(\Delta\varphi_0) \simeq \Delta\varphi_0 . \quad (30)$$

With the help of Eqs.(28) and (29), Eq.(27) may be written as

$$W \simeq \frac{m}{h^2}\left[1 + e\cos(\varphi - \varphi_0 - \Delta\varphi_0)\right] . \quad (31)$$

From Eq.(31) we can see that while the planet moves through an angle φ, the perihelion advances for a fraction of the revolution angle equal to

$$\frac{\Delta\varphi_0}{\varphi} = \frac{3m^2}{h^2} = \frac{3m}{a(1 - e^2)} . \quad (32)$$

To get the far right result, we have made use of Eq.(22). After a full revolution $\varphi = 2\pi$, and we have

$$\Delta\varphi_0 = \frac{6\pi m}{a(1 - e^2)} = \frac{6\pi GM}{c^2 a(1 - e^2)} . \quad (33)$$

The dependence of the perihelion advance on the eccentricity e and the semimajor axis a is evident. In Table 1, we list observational data for

Mercury, Venus, Earth and Icarus (asteroid); their semimajor axes are small enough and M large enough for $\Delta\varphi_0$ to be measured. Shown is the perihelion advance per century.

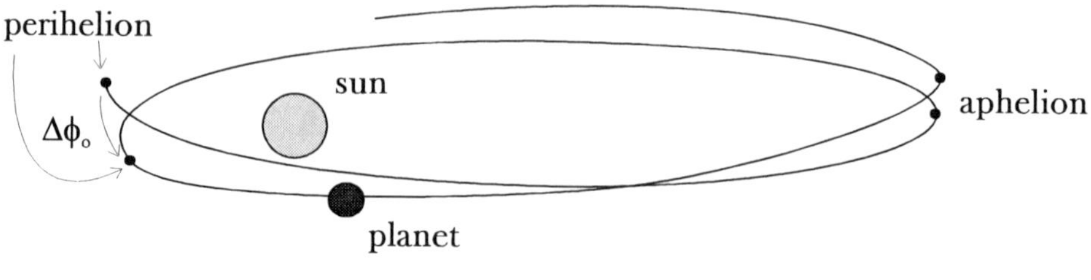

Figure 1: The shift of the perihelion.

Table 1: Observational data for selected planets

Planet	a(10^6 km)	e	φ (seconds of arc per century) Observed	Theoretical
Mercury	57.91	0.2056	43.11 ± 0.45	43.03
Venus	108.21	0.0068	8.4 ± 4.8	8.6
Earth	149.60	0.0167	5.0 ± 1.2	3.8
Icarus	161.0	0.827	9.8 ± 0.8	10.3

The precision of these observations is quite remarkable when we consider that for the planet Mercury, the measured perihelion shift per century amounts to

$$\text{(Mercury)} = 5600.73 \pm 0.41''$$

Of this value, the shift caused by known nonrelativistic disturbance effects is

$$\text{(disturbance)} = 5557.62 \pm 0.20''$$

43.11 ± 0.45" is attributed to relativistic effect. The advance of Mercury's perihelion was known to astronomers as early as 1860. Its final explanation through the general theory of relativity in 1915 was one of the greatest triumph of Einstein's theory.

7-2 Deflection of light rays

We saw in Chapter 3 that Einstein's principle of equivalence implies that light is deflected in a gravitational field. To investigate the deflection of light in a gravitational field, we can regard a light beam composed of a large number of photons, each one moving with a speed c and with vanishing rest mass. Such a particle follows a null geodesic, $ds^2 = 0$. We need to again solve the geodesic equations given by Eqs.(2), (3), (4), and (5). The two integrals (10) and (11) are still valid. But instead of Eq.(12), we obtain a corresponding relation for null intervals. We first set the Schwarzschild squared-line element equal to zero, and then divide through by ds^2:

$$e^{-2\nu}\left(\frac{dr}{ds}\right)^2 + r^2\left(\frac{d\varphi}{ds}\right)^2 - e^{2\nu}c^2\left(\frac{dt}{ds}\right)^2 = 0\,. \qquad (34)$$

The quantity s is now a suitable variable which parameterizes the null geodesic. Employing Eqs.(10) and (11) in Eq.(34) and letting $W = 1/r$ leads to the equation

$$\left(\frac{dW}{d\varphi}\right)^2 = \left(\frac{ck}{h}\right)^2 - W^2\left(1 - \frac{2GM}{c^2}W\right) \qquad (35)$$

from which it follows, by differentiating

$$\frac{d^2W}{d\varphi^2} + W = \frac{3GM}{c^2}W^2 \qquad (36)$$

which is much easier to deal with. As before the term $(3GM/c^2)W^2$ is small in comparison with other terms, so that we can get an approximate solution by an iterative procedure, starting with a zero-order equation

$$\frac{d^2W^{(0)}}{d\varphi^2} + W^{(0)} = 0 \tag{37}$$

which has the solution

$$W^{(0)} = \frac{1}{R}\cos(\varphi - \varphi_0) \tag{38}$$

where φ_0 and R are integration constants. It is a straight path, with R as the distance of closest approach for the unperturbed light ray. In zero-order approximation, the light ray is not deflected by the gravitational field of the gravitating mass. In a successive approximation we have

$$\frac{d^2W^{(1)}}{d\varphi^2} + W^{(1)} = 3mW^{(0)} = \frac{3m}{R}\cos^2\varphi \tag{39}$$

where $m = GM/c^2$. This equation will not lead to a straight path and a particular integral of this equation is

$$W^{(1)} = \frac{m}{R^2}(1 + \sin^2\varphi) \tag{40}$$

so that the trajectory, to first order, is given by

$$W \simeq W^{(0)} + W^{(1)} = \frac{\cos\varphi}{R} + \frac{m}{R}(1 + \sin^2\varphi) \; . \tag{41}$$

We are ready to calculate the angle, found between the two asymptotes, to the curve described by the above equation. To get the asymptote we have to let $r \to \infty$ (W = 0) and we have

$$\cos\varphi + \alpha(1 + \sin^2\varphi) = 0; \qquad \alpha = \frac{m}{R} = \frac{GM}{c^2R}$$

or

$$\cos\varphi + \alpha(2 - \cos^2\varphi) = 0 \; . \tag{42}$$

The solution to this quadratic equation is

$$\cos\varphi = \frac{1 - \sqrt{1 + 8\alpha^2}}{2}\alpha \simeq -2\alpha = -\frac{2GM}{c^2R} \; . \tag{43}$$

GM/c^2R is very small, so φ is very close to $\pi/2$. Setting, accordingly,

$$\varphi = \frac{\pi}{2} + \delta \tag{44}$$

we get for one asymptote

$$\sin\delta \simeq \delta = \frac{2GM}{c^2R} ; \tag{45}$$

For the other asymptote we take

$$\varphi = -\frac{\pi}{2} - \delta \tag{46}$$

and this leads to the same value for $\delta(= 2GM/c^2R)$. Thus the angle between the asymptotes is

$$\Delta\varphi = 2\delta = \frac{4GM}{c^2R} . \tag{47}$$

This is the total deflection of a light ray in the gravitational field of the mass M. It is important to note that this angle is just twice the value predicted by the Newtonian theory of gravitation.

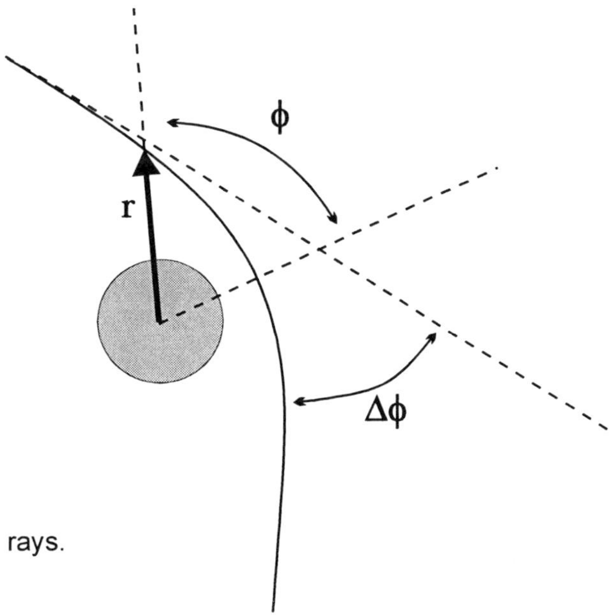

Figure 2: Deflection of light rays.

A light ray coming from a distant star undergoes a maximum deflection when it grazes the sun, i.e. when the distance R is just the radius of the sun, $R = 7 \times 10^8$ m: $\Delta\varphi = 4GM/c^2R = 1.75''$. For the deflection as observed during eclipses, the individual measured values have a probable error of roughly 10 percent and are usually within 10 percent of the value given by Eq.(47). The average for eleven eclipses agrees with Eq.(47) to roughly one part in 500.

7-3 Radar echoes from planets (the Shapiro experiment)

In 1965, I. I. Shapiro proposed a new test of general relativity in the environment of our solar system, which become possible through progress made in radar technology and the design of more accurate atomic clocks. The basic idea of Shapiro's experiment is simple: a radar signal is sent from earth toward a planet, say, Venus, where it is reflected and arrives back at earth. The time it takes to travel out and back is accurately measured. If the signal has to pass near the sun's edge on its way from earth to Venus and back, the echo would be delayed relative to an imaginary signal which travels out and back in a straight line with speed c. The slowdown of clocks and deflection of radar signal in sun's gravitational field contribute to this effect. To see this, we use the Schwarzschild metric to calculate the coordinate speed of radar signal travelling in the (r, φ) plane defined by $\theta = \pi/2$. For radar signal, $ds = 0$:

$$0 = c^2\left(1 - \frac{2GM}{c^2r}\right)dt^2 - r^2d\varphi^2 - \left(1 - \frac{2GM}{c^2r}\right)^{-1}dr^2 \tag{48}$$

from which we obtain

$$0 = 1 - \frac{2GM}{c^2r} - \frac{1}{c^2}\left[\frac{\left(\frac{dr}{dt}\right)^2}{1 - \frac{2GM}{c^2r}} + r^2\left(\frac{d\varphi}{dt}\right)^2\right]. \tag{49}$$

To eliminate $d\varphi/dt$ we use the two integrals (Eqs.(10) and (11))

$$r^2\frac{d\varphi}{ds} = h \qquad \left(1 - \frac{2GM}{c^2r}\right)\frac{dt}{ds} = k \ .$$

Their quotient gives

$$\frac{r^2 d\varphi/ds}{\left(1 - \frac{2GM}{c^2r}\right)dt/ds} = \frac{r^2 d\varphi/dt}{1 - \frac{2GM}{c^2r}} = \frac{h}{k} \equiv D \ .$$

Thus Eq.(49) becomes

$$0 = 1 - \frac{2GM}{c^2r} - \frac{1}{c^2}\left[\frac{(dr/dt)^2}{1 - \frac{2GM}{c^2r}} + \frac{D^2}{r^2}\left(1 - \frac{2GM}{c^2r}\right)\right] . \tag{50}$$

We can relate D to the closest approach distance r_0 by using the fact that the radar signal has no radial component at the closest approach point where $r = r_0$. This means that $dr/dt = 0$ at $r = r_0$. Then Eq.(50) gives

$$\frac{dr}{dt} = c\left(1 - \frac{2GM}{c^2r}\right)\sqrt{1 - \frac{r_0^2\left(1 - \frac{2GM}{c^2r}\right)}{r^2\left(1 - \frac{2GM}{c^2r_0}\right)}} \tag{51}$$

By integration, we obtain

$$t(r,r_0) = \frac{1}{c}\int_{r_0}^{r} \frac{dr}{\left(1 - \frac{2GM}{c^2r}\right)\sqrt{1 - \left[r_0^2\left(1 - \frac{2GM}{c^2r}\right)/r^2\left(1 - \frac{2GM}{c^2r_0}\right)\right]}} \tag{52}$$

where $t(r,r_0)$ denotes the coordinate time taken for the signal to travel from r in to the radius of closest approach (Figure 3). If the signal travelled in straight lines with speed c, we would have $t(r,r_0) = t^0(r,r_0)$ where

$$t^0(r,r_0) = \sqrt{\frac{r^2 - r_0^2}{c}} \; . \tag{53}$$

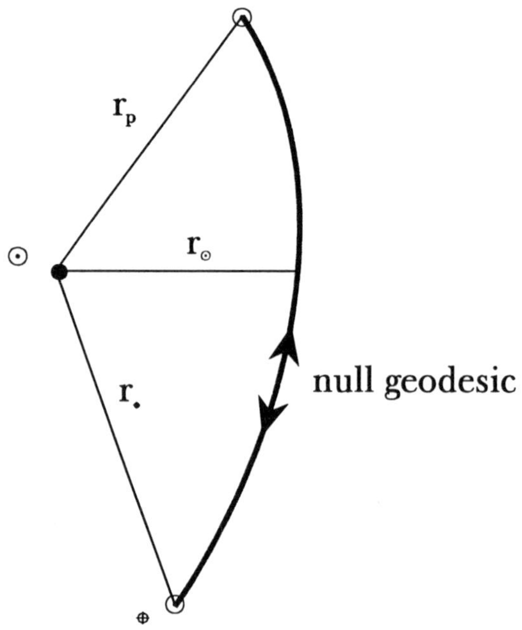

Figure 3: Radar time delay of signals.

Let us go back to Eq.(52) and work out the integral to first order in the small quantity GM/c^2r. The expansion of the integrand is not quite straightforward, but leads to the result

$$t(r,r_0) \simeq \frac{1}{c}\int_{r_0}^{r} \frac{r\,dr}{\sqrt{r^2 - r_0^2}}\left[1 + \frac{2GM}{c^2 r} + \frac{GMr_0}{c^2 r(r + r_0)}\right] . \tag{54}$$

The integrals are elementary, and give

$$t(r,r_0) = \frac{1}{c}\left[\sqrt{r^2 - r_0^2} + \frac{2GM}{c^2}\ln\left(\frac{r + \sqrt{r^2 - r_0^2}}{r_0}\right) + \frac{GM}{c^2}\sqrt{\frac{r - r_0}{r + r_0}}\right] . \tag{55}$$

It is seen how the terms in GM/c^2 provide a small correction to $t^0(r,r_0)$. The coordinate time T between the emission of the signal and the reception of the echo is

$$T = 2\left[t(r_\oplus, r_0) + t(r_p, r_0)\right] \tag{56}$$

where $r_\oplus$ and r_p are the coordinate distance of the earth and planet from the sun. T is the coordinate time between two events on the earth, that is, at essentially fixed r, θ, φ. The excess time delay ΔT is defined as

$$\begin{aligned} \Delta T &= T - \frac{2}{c}\left[t^0(r_\oplus, r_0) + t^0(r_p, r_0)\right] \\ &= T - \frac{2}{c}\sqrt{r_\oplus^2 - r_0^2} - \frac{2}{c}\sqrt{r_p^2 - r_0^2} \quad . \end{aligned} \tag{57}$$

In practical situations r_0 is approximately the sun's radius, so $r_\oplus >> r_0$ and $r_p >> r_0$. Therefore the square roots terms in Eq.(55) can be simplified, and the final result is

$$\Delta T \simeq \frac{4GM_\oplus}{c^3}\left[\ln\left(\frac{4r_\oplus r_p}{r_0^2}\right) + 1\right] . \tag{58}$$

For an echo from Venus which just grazes the sun ($r_0 = r_\odot$), $T = 2.4 \times 10^{-4}$ seconds, which is easily measured. Several experiments have been performed by Shapiro and his coworkers; in every case the predictions of general relativity are confirmed (with an uncertainty of 3%). Figure 4 is the result for the superior conjunctions of Venus in 1970.

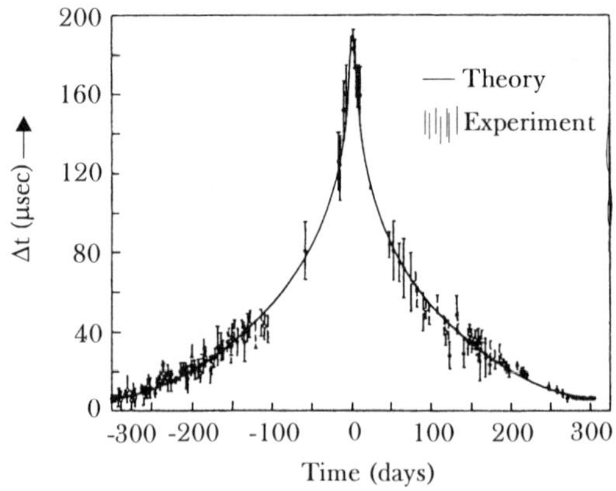

Figure 4: Radar time delay of signals from Venus: Comparison of theory and experiment.

7-4 The Hafele-Keating experiment

The Hafele-Keating experiment was designed to measure the influence of the earth's gravitational field on clocks. A few decades ago, this was an impossible task. However, by 1960 the precision of caesium clocks was improved to such a degree that even subtle irregularities in the earth's rotation could be detected.

Consider two identical atomic clocks A and B, synchronized at rest on earth's surface. Clock A remains at rest, while clock B is flown around the world at height h in a airplane whose speed relative to the ground is v (Figure 5). Since clock B is at a lower gravitational potential, it will go

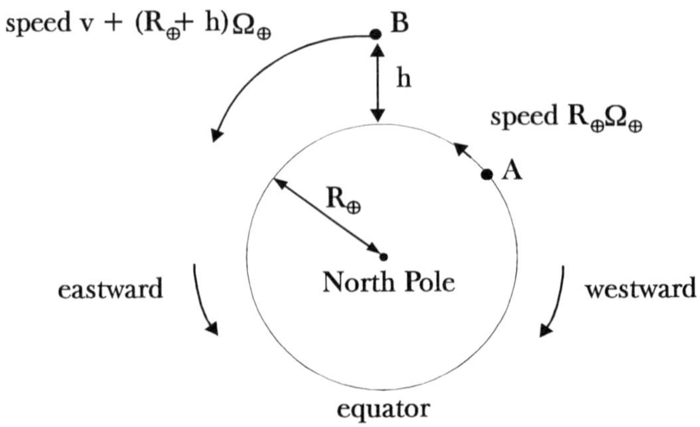

Figure 5: Around the world with clocks.

faster than clock A. After once around the world, the proper time τ_B experienced by clock B is compared with the proper time τ_A experienced by clock A. To calculate τ_A, τ_B we use the Schwarzschild metric again, assuming for simplicity that the circumnavigation is equatorial and remembering that the earth is rotating with angular velocity $\Omega_\oplus$ relative to the local inertial frame in which it is embedded (i.e. a frame in free fall in the sun's field). For clock A, r = R, $\Delta r = 0$, $\Delta\theta = 0$, and $\theta = \pi/2$. Thus

$$\tau_A = \int \Delta\tau_A = \int \sqrt{\left(1 - \frac{2GM_\oplus}{c^2 R_\oplus}\right)\Delta t^2 - R_\oplus^2 \Delta\frac{\varphi^2}{c^2}} \quad . \tag{59}$$

Now the speed of clock A relative to the local inertial frame is

$$R_\oplus \Omega_\oplus = R_\oplus \frac{d\varphi}{dt} \tag{60}$$

so that

$$\begin{aligned} \tau_A &= \int_0^t dt \sqrt{1 - \frac{2GM_\oplus}{c^2 R_\oplus} - R_\oplus^2 \frac{\Omega_\oplus^2}{c^2}} \\ &= t\sqrt{1 - \frac{2GM_\oplus}{c^2 R_\oplus} - \frac{R_\oplus^2 \Omega_\oplus^2}{c^2}} \end{aligned} \tag{61}$$

where t is the coordinate time between the initial synchronization event I and the final comparison event F. For B we neglect the small effects of rising to, and descending from, the height h. So

$$\tau_B = \int \sqrt{1 - \frac{2GM_\oplus}{c^2(R_\oplus + h)} - \frac{(R_\oplus + h)(\Delta\varphi/\Delta t)^2}{c^2}} \quad . \tag{62}$$

The coordinate speed of clock B is

$$(R_\oplus + h)(\Delta\varphi/\Delta t) \simeq (R_\oplus + h)\Omega_\oplus + v, \tag{63}$$

where the equality is approximate because we have neglected the relativistic correction to the velocity addition formula. A positive v refers to eastward motion (i.e. motion with the earth). Thus

$$\tau_B = t\sqrt{1 - \frac{2GM_\oplus}{c^2(R_\oplus + h)} - \frac{[(R_\oplus + h)\Omega_\oplus + v]^2}{c^2}} \tag{64}$$

where t is again the coordinate time between events I and F. To eliminate the coordinate time t, we can calculate the "time offset" δ, defined by

$$\delta \equiv \frac{\tau_B - \tau_A}{\tau_A} . \tag{65}$$

We can further simplify the square roots, because

$$\frac{GM_\oplus}{c^2 R_\oplus} \ll 1, \quad \frac{h}{R_\oplus} \ll 1, \quad \frac{v^2}{c^2} \ll 1, \quad \frac{h\Omega_\oplus}{v} \ll 1 . \tag{66}$$

To the lowest order in these four quantities, we get

$$\delta = \frac{gh}{c^2} - \frac{(2R_\oplus\Omega_\oplus + v)}{2} c^2 \tag{67}$$

where

$$g = GM_\oplus / R_\oplus^2 = 9.81 \text{ m s}^{-2} .$$

If we take $h = 10^4$ m and a typical jet speed $v = 300$ m s^{-1}, we obtain

$$\delta_{\text{westward}(v<0)} = 2.1 \times 10^{-12} \qquad \delta_{\text{eastward}(v>0)} = -1.0 \times 10^{-12} . \tag{68}$$

The time offset is very small, but it is observable; the fractional accuracy of modern caesium clocks is about 10^{-13}. In 1971 Keating and Hafele made eastward and westward flights round the world on commercial jets, carrying caesium clocks which they later compared with identical caesium clocks at the U.S. Naval Observatory in Washington, D.C. Because the flights were not equatorial, they employed a variant of formula (67) for δ. But the theory is essentially the same, and their results (Table 2) give a direct test of the dependence of the separation between two events on the

Table 2: Hafele-Keating experiment

Direction of flight	(nanoseconds) Experiment	Theory
Westerward	273 ± 7	275 ± 21
Easterward	-59 ± 10	-40 ± 23

world line joining them. Their results also successfully test the Schwarzschild metric (6-26). It is interesting to note that the Hafele-Keating results conclusively resolve the "twin paradox" controversy in Einstein's favor.

The four experiments discussed in this chapter constitute almost the entire experimental verification of Einstein's general relativity. This is in sharp contrast with the quantum theory which is supported by an abundance of experimental data. The main problem in designing a feasible experimental test of general relativity is to some extent a technological one. To observe the very small effects that the very weak interaction of the theory predicted requires sophisticated techniques which are usually not available.

Problems

(7-1) Verify the results of Eqs. (2) to (5).

(7-2) Calculate the radial coordinate at which light travels in a circular path round a body of mass M, using (a) Newtonian mechanics, (b) general relativity. Express the answers as a multiple of the Schwarzschild radius.

(7-3) Considering the light-photon as a projectile moving under the influence of Newtonian gravity (inverse square law), calculate the bending of light produced by a massive body. Show that the net bending is half that given by general relativity.

Chapter 8

The Physics of Black Holes

8-1 The Schwarzschild black hole

We noticed in Chapter 6 that in the Schwarzschild solution the metric coefficients g_{11} (i.e. g_{rr}) become infinite if r equals the gravitational radius r_g (also known as the Schwarzschild radius r_s) defined as

$$r_g = 2GM/c^2$$

and a black hole is formed. Matter may fall into black holes but nothing (including light signals) can come out of a black hole. We must now prove the truth of these startling assertions. To this aim let us consider a particle falling radially into the central body and let its velocity vector be $v^1 = dx/ds$. Since the particle falls in radially, we can take $v^2 = v^3 = 0$. The motion is described by the geodesic equation

$$\frac{dv^{\mu}}{ds} + \Gamma^{\mu}_{\nu\sigma} v^{\nu} v^{\sigma} = 0 \tag{1}$$

which reduces to, for the case we are considering

$$\frac{dv^0}{ds} = -\Gamma^0_{\mu\nu} v^{\mu} v^{\nu} = -g^{00}\Gamma_{0,\mu\nu} v^{\mu} v^{\nu} = -2g^{00}\Gamma_{0,10} v^0 v^1 .$$

From

$$\Gamma_{\mu,\nu\sigma} = \frac{1}{2}(g_{\mu\nu,\sigma} + g_{\mu\sigma,\nu} - g_{\nu\sigma,\mu})$$

we find

$$\Gamma_{0,10} = \frac{1}{2} g_{00,1} = \frac{1}{2}\frac{\partial g_{00}}{\partial x^1}$$

so Eq.(1) becomes

$$\frac{dv^0}{ds} = -g^{00}\partial g_{00,1}v^0\frac{dx^1}{ds} = -g^{00}\frac{dg_{00}}{ds}v^0 .$$

Now $g^{00} = 1/g_{00}$, so we finally get

$$g_{00}\frac{dv^0}{ds} + \frac{dg_{00}}{ds}v^0 = \frac{d}{ds}\left(g_{00}v^0\right) = 0 .$$

This integrates to

$$g_{00}v^0 = k, \tag{2}$$

with k an integration constant (it is the value of g_{00} where the particle starts to fall). And from

$$ds^2 = g_{\mu\nu}dx^\mu dx^\nu$$

we have

$$1 = g_{\mu\nu}v^\mu v^\nu = g_{00}(v^0)^2 + g_{11}(v^1)^2 .$$

Multiplying this equation by g_{00} we obtain

$$g_{00} = (g_{00})^2(v^0)^2 + g_{00}g_{11}(v^1)^2 . \tag{3}$$

Now from Chapter 6, we have

$$g_{00}g_{11} = -e^{2(\lambda+\nu)} = -e^0 = -1 .$$

Substituting this and Eq.(2) into Eq.(3) we get

$$k^2 - (v^1)^2 = g_{00} = 1 - r_g/r$$

from which we obtain

$$(v^1)^2 = k^2 - 1 + r_g/r . \tag{4}$$

For a falling body $v^1 < 0$, and hence

$$v^1 = -\sqrt{k^2 - 1 + r_g/r} . \tag{4a}$$

Now let us consider dt/dr

$$\frac{dt}{dr} = \frac{dx^0/ds}{dx^1/ds} = \frac{v^0}{v^1}$$

and from Eq.(2) we have

$$v^0 = k/g_{00} = k(1 - r_g/r)$$

so

$$dt/dr = v^0/v^1 = -k(1 - r_g/r)^{-1}(k^2 - 1 + r_g/r)^{-1/2} . \qquad (5)$$

Let us now suppose the particle is close to the critical radius r_g, so $r = r_g + \epsilon$, with ϵ small, and let us neglect ϵ^2. Then

$$\begin{aligned} dt/dr &= -k[1 - (1 + \epsilon/r_g)^{-1})^{-1}[k^2 - 1 + (1 + \epsilon/r_g)^{-1}]^{-1/2} \\ &= - k\ (\epsilon/r_g)^{-1}(k^2 + \epsilon/r_g)^{-1/2} = - (\epsilon/r_g)^{-1} \end{aligned}$$

or

$$dt/dr = - r_g/\epsilon = - r_g/(r - r_g). \qquad (6)$$

This integrates to

$$t = - r_g \log(r - r_g) + \text{const.} \qquad (7)$$

Thus, as $r \rightarrow r_g$, $t \rightarrow \infty$. The particles takes an infinite time to reach the Schwarzschild radius $r = r_g = 2GM/c^2$.

We now consider an adventurer travelling with the particle. His time is measured by ds. Now

$$ds/dr = 1/v^1 = - (k^2 - 1 + r_g/r)^{-1/2} \qquad (8)$$

and this tends to $- k^{-1}$ as r tends to r_g. Thus the particle and the adventurer reach $r = r_g$ after the lapse of finite proper time for them. The singularity at $r = r_g$ is therefore not a real unphysical singularity, it is only a coordinate singularity due to the choice of coordinate systems. We shall come back to this point later. But odd things do happen at $r = r_g$. If the adventurer signals to a distant observer by sending light flashes at intervals which are precisely regular according to his (the adventurer's) proper time, the light is red-shifted by a factor $g_{00}^{-1/2} = (1 - r_g/r)^{-1/2}$ as received by the distant observer. This factor becomes infinite as the adventurer approaches the gravitational radius r_g. Also, according to the distant observer, the time intervals between the received light flashes become longer and longer as the adventurer approaches r_g. Thus, an adventurer falling radially inward appears to continue beyond the threshold at $r = r_g$, and a distant observer viewing his fall only sees him before he passes the threshold at $r = r_g$. The lack of communication with the outside world (through light or other material signals) is a basic property of black holes. The boundary $r = r_g$ is called an **event horizon**. Just as the curved earth leads to the existence of a horizon that limits the range of vision of a navigator on the high seas, the curved geometry of the black holes also leads to the formation of an event horizon which hides from the external observer of spacetime its interior. However, black holes still exert influence on their surroundings, because their gravitational effects on external bodies arise from the Schwarzschild metric for $r > r_g$.

8-2 Inside a black hole

A black hole has no exact Newtonian analogue, and its many strange properties can only be described properly with Einstein's theory of general relativity. We now discuss some of these strange properties.

First, we shall see that there is no static inside a black hole, only dynamics. The argument runs as follows: first we recall that the world line of any particle must be time-like, i.e. the line interval ds^2 along the world line is always positive. Now for a particle at rest, we have

$$dr = d\theta = d\varphi = 0$$

but inside the Schwarzschild sphere ($r < r_g$), the coefficient g_{00} ($= 1 - r_g/r$) of dt^2 is negative, so that ds^2

$$ds^2 = c^2 d\tau^2 = c^2(1 - r_g/r)dt^2$$

is negative and it is not time-like. Thus, there is no stationary particle inside a black hole, only dynamics. Then, how do particles move inside a black hole? Do particles move inward or outward? To answer these questions, we examine the continuity of <u>null cones</u>. These are the double cones joining an event P to those neighboring events corresponding to zero separation. All possible world lines for which $d\tau^2 > 0$ (the time-like geodesics from P) all lie within the cones. The lines in one cone point into P's future, and those in the other cone point into P's past. Since the coordinates (t, r, θ, φ) are inadequate for discussing what happens at $r < r_g$, we introduce a new time coordinate which is not singular at r_g:

$$t' = t + (2GM/c^2)\ln|rc^2/2GM - 1| \tag{9}$$

from which we obtain

$$dt = dt' - (r_g/cr)(1 - r_g/r)^{-1}dr \tag{10}$$

and

$$dt^2 = dt'^2 - (2r_g/cr)(1 - r_g/r)^{-1}dt'dr + (r_g/cr)^2(1 - r_g/r)^{-2}dr^2 \,. \tag{11}$$

For a null cone we have

$$(1 - r_g/r)dt^2 - c^{-2}(1 - r_g/r)dr^2 = 0 \,.$$

Combining this with Eq.(11) we obtain

$$(1 - r_g/r)dt'^2 - (2r_g/cr)drdt' - (dr^2/c^2)(1 + r_g/r) = 0 \,. \tag{12}$$

Eq.(12) is a quadratic equation in dt'^2 which has 2 roots given by

$$dt'/dr = -\frac{1}{c} \,; \qquad dt'/dr = c^{-1}[(1 + r_g/r)/(1 - r_g/r)] \,. \tag{13}$$

For $r > r_g$ the future null cones point upwards, that is, the world lines correspond to increasing coordinate time. Some lines in the future cones point inward, and some outward, that is, particles can travel either toward or away from the central mass. As we approach the black hole, the future cones lean inward until at $r < r_g$ all world lines point inward towards r = 0. Thus, anything (matter and radiation) inside a black hole must fall into the center r = 0. Figure 1 shows light cones in the plane of r and the new time coordinate t′. A photon starting where $r > r_g$ can travel inward, cross the threshold at $r = r_g$ and carry on inward, but a photon

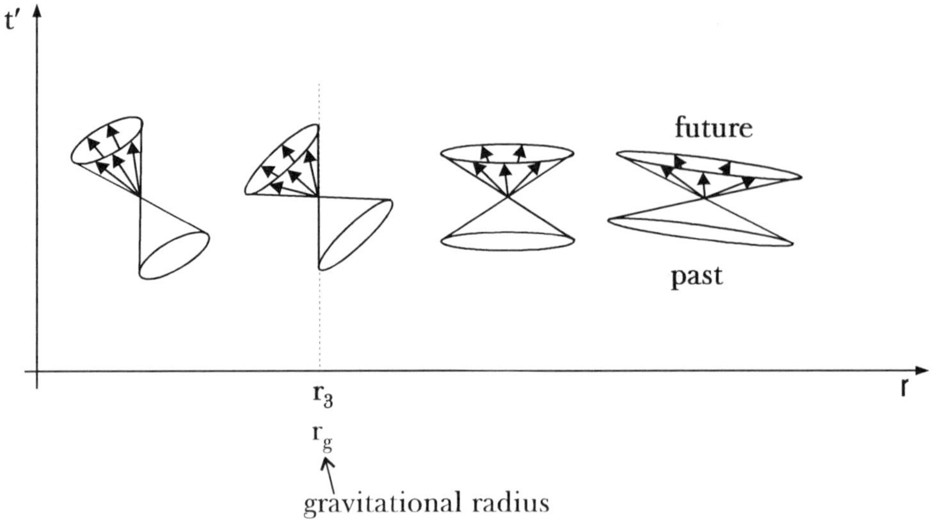

Figure 1: Light cones near a black hole.

starting where $r < r_g$ does not travel outward. Thus, an outside observer could only detect the presence of a black hole through its gravitational field.

8-3 How a black hole may form

Our discussion of the properties of a black hole would be academic, unless there were reasons for believing that they might exist in nature. The possibility of their existence arises from the idea of gravitational collapse, first studied by Oppenheimer and Volkoff in 1939. Astrophysicists have calculated that the ultimate stage in the evolution of massive stars ($M > 10$ $M_\odot$, where $M_\odot$ is the solar mass) would be gravitational collapse. In the process of collapse, a substantial fraction of the mass will be returned to the interstellar medium. If the mass ejected is such that what remains is in the permissible range of masses for a stable neutron star, then a pulsar will be formed. The current estimate of the permissible range of masses for stable neutron stars is between 0.3 to 2.0 $M_{\odot}$. The exact specification of the permissible range of masses for stable neutron stars depends on the equation of state for neutron matter. If the star ejects an amount that is either too large or too little, the residue will not be able to settle into a stable neutron star state, and the process of collapse must continue until the gravitational force becomes so strong that a black hole is formed.

Black holes are very strange objects. Their most astonishing property is that, as we saw in the previous section, anything inside the Schwarzschild sphere ($r < r_g$) must fall into the center. This applies to the matter constituting the star whose collapse formed the black hole. This implies that the collapse continues until the star is a point singularity at $r = 0$. Of course, the implication holds only if general relativity remains valid in the unimaginably dense, hot conditions near the end point of collapse. It has been argued that some new force may come into play to prevent the ultimate formation of a true singularity. For the present this is a matter of speculation. Quantum effects are not expected to become dominant with gravity until we are dealing with very short distances, possibly as short as the Planck length, 10^{-35} m. However, no consistent theory of quantum gravity has yet been constructed, so this is still very much an open question.

In our universe we may find black holes of masses ranging between 2 and 3 solar masses or even more, resulting from stellar collapses. Supermassive black holes, containing thousands, millions or billions of solar masses may also exist. It has also been suggested that, if our universe began in a hot dense Big Bang, conditions in the earliest moments may have been such that quite small amounts of material could have been squeezed sufficiently to form "mini black holes".

Figure 2 presents the gravitational collapse in a spacetime diagram, from the collapse to the development of a black hole (from bottom to top). Shown is the collapse as the center of the star, illustrated here in a circular cross section. The vertical line in the center is the world line of the star's center. As we move upward (so forward in time) we see that the world line is surrounded by circles of ever decreasing radii. These are the cross-sectional disks that collapse with time. For an observer on the surface of the collapsing star (observer 1), nothing unusual happens as he crosses r_g (the Schwarzschild surface) at the time

$$\tau_{(r=r_g)} = -2r_g/3c \ . \tag{14}$$

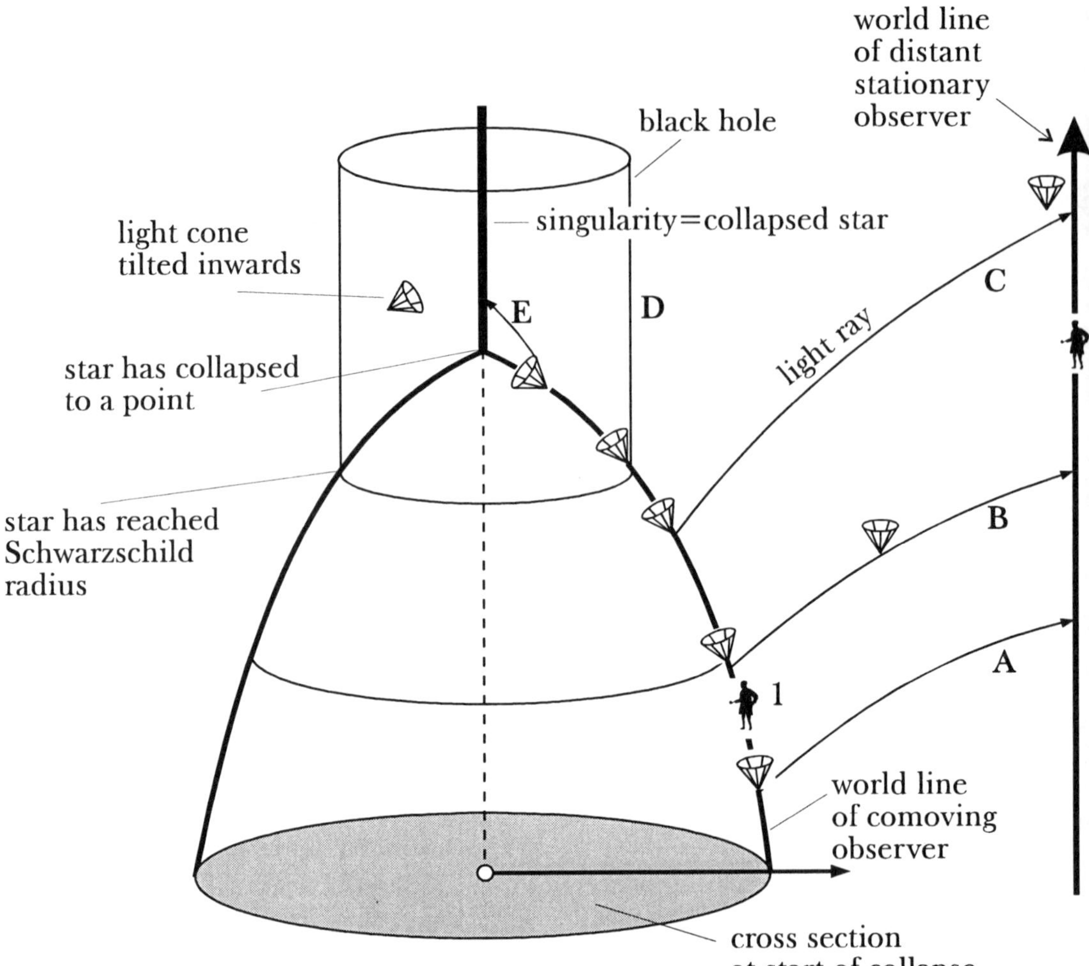

Figure 2: Space-time diagram of a collapsing star.

It is easy to derive the above formula. We have already worked out the necessary equation in section 1, and let us recall it:

$$(v^1)^2 = k^2 - 1 + r_g/r \ ; \qquad v^1 = \frac{dr}{ds}$$

Initially (i.e. before the star starts to collapse) v^1 is zero and the ratio r_g/r is negligibly small. This implies that $k^2 - 1 = 0$, so that

$$\left(\frac{dr}{ds}\right)^2 = \frac{r_g}{r}$$

or

$$\sqrt{r}dr = \sqrt{2GM}\,d\tau \ .$$

If the travelling clock (i.e. the clock on the surface of the collapsing star) is set so as to read $\tau = 0$ at the moment of collapse at $r = 0$, then the solution of this equation is

$$r(\tau) = (-3\tau/2)^{3/2}(2GM)^{1/3} \qquad (15)$$

and the observer on the surface of the collapsing star crosses r_g at the time given by Eq.(8). After crossing r_g the collapse is very rapid: for a black hole of solar mass it only takes about 10^{-5} sec to reach the singularity at $r = 0$.

The story is quite different for an observer (2) who watches the collapsing of the star at a safe and constant distance from the star. Suppose that observer (1) communicates with observer (2) during the collapse, by sending out light signals at constant time intervals (measured in his own time) toward observer 2. Those signals are labelled A, B, C, D, and E and are directed radially away from the star's surface. Although the signals are sent out by observer (1) at precisely equal time intervals, their arrivals at observer (2) are gradually delayed as the collapse progresses. This is easy to see. If $\Delta\tau_E$ is the proper time interval of observer (1) who sends out signals from an emitter and $\Delta\tau_R$ the proper time interval of observer (2) who receives the signals, then we have (see preceding chapter).

$$\Delta\tau_R = \sqrt{\frac{1 - r_g/r_R}{1 - r_g/r_E}}\,\Delta\tau_E \ . \qquad (16)$$

Thus, signals A and B arrive at observer (2) at about the same time interval with which they departed from observer (1), and signal C arrives considerably delayed due to the increased influence of the gravitational field. Just as observer (1) crosses r_g, he sends signal D, which never arrives at observer (2), it is trapped at $r = r_g$ (vertical edge of the light cone). The last signal, E, will quickly fall into the singularity at $r = 0$. This discussion demonstrates that the collapse when observed from a distance will appear to decrease gradually until it stops entirely, or appears "frozen" at the $r = r_g$. Black holes are therefore often called **frozen stars.**

The luminosity of the collapsing star also decreases rapidly since the light will be more and more red-shifted, the closer to the gravitational radius r_g it is emitted. A further reduction in the luminosity results from the fact that the photons emitted at equal time intervals at observer (1) will reach observer (2) at ever increasing time intervals, thus the total number of photons received per unit time is decreasing as the collapse progresses. Detailed calculations reveal that the luminosity L of the star during the last phase of the collapse near r_g diminishes exponentially:

$$L = \text{const.} \times e^{-ct/r_g} \,. \tag{17}$$

Thus, for all practical purposes a collapsing star does appear to switch off like a light.

8-4 Other coordinate systems

The description of spacetime near a spherically symmetric massive object need not be in terms the standard Schwarzschild coordinates and their corresponding line element. In this section we introduce two more coordinate systems: the isotropic coordinates and the Kruskal coordinates.

(A) The isotropic coordinates

Here the standard coordinate r is replaced by ρ, defined by

$$r \equiv \rho\left[1 + r_g/(4\rho)\right]^2 \tag{18}$$

Simple substitution gives the line element

$$ds^2 = c^2(1 - r_g/4\rho)^2(1 + r_g/4\rho)^{-2} - (1 + r_g/4\rho)^4(d\rho^2 + \rho^2 d\theta^2 + \varphi^2 \sin^2\theta d\varphi^2) \ . \quad (19)$$

It is easy to see that there is no coordinate singularity. The isotropic line element has the form

$$ds^2 = A(\rho)dt^2 + B(\rho)d\sigma^2 \quad (20)$$

where $d\sigma^2$ is the line element of flat spacetime in spherical polar coordinates ρ, θ, φ. The advantage of the isotropic line element is that $d\sigma^2$ is invariant under changes of flat-space coordinates, and so ρ, θ, φ may be replaced by any other flat-space coordinates we care to use. For example, if cartesian coordinates x, y, z (defined in terms of ρ, θ, φ in the usual way) are used as spatial coordinates, then

$$d\sigma^2 = dx^2 + dy^2 + dz^2 \quad (21)$$

and

$$ds^2 = A(\rho)dt^2 + B(\rho)(dx^2 + dy^2 + dz^2) \quad (22)$$

where

$$\rho 2 = dx^2 + dy^2 + dz^2 \ .$$

Our previous results could be expressed in terms of isotropic coordinates, but the corresponding expressions are usually much more complicated. However, these coordinates are employed in compiling relativistic astronomical tables for the solar system, and are so used extensively throughout the world.

(B) Kruskal coordinates

Kruskal coordinates are particularly useful for discussing events on both sides of the event horizon. The r and t of the standard Schwarzschild coordinates are now replaced by

$$q^2 = (r/r_g - 1)e^{r/r_g}\cosh(ct/2r_g) \tag{23}$$

$$p^2 = (r/r_g - 1)e^{r/r_g}\sinh(ct/2r_g) \ . \tag{24}$$

This leads to the line element

$$ds^2 = (4r_g/r)e^{-r/r_g}(dq^2 - dp^2) + r^2 d\Omega^2 \tag{25}$$

where

$$d\Omega^2 = d\theta^2 + \sin^2 d\varphi^2 \ . \tag{26}$$

The particular advantage of these coordinates is that the radial null geodesics are given by $q^2 \pm p^2$ = constant, and are thus straight lines with 45° slopes when drawn in a q,p-diagram, just as the flat spacetime of special relativity.

Note that no amount of coordinate transformation from one system to another can change the curvature of the spacetime. That is, we cannot transform away gravity just by turning to another coordinate system, except locally, but then only approximately.

8-5 The Kerr-Newman black hole

The Schwarzschild black holes we have discussed so far are of a very special kind: they are non-rotating. However, rotation is a property common to stars, planets and galaxies alike. A rotating star possesses angular momentum, and during the collapse of such a rotating star, we may expect the angular momentum to be retained except for the part which may be radiated away in gravitational waves. If a black hole were to form as a result of the collapse of a rotating star, we would expect the black hole to be rotating at a rapid rate; after all, we know that neutron stars spin very rapidly indeed.

As a result of work carried out by R. H. Price, B. Carter, W. Israel, D. C. Robinson and S. W. Hawking, it has been shown that black holes, from the point of view of the outside observer, can possess only three distinguishing characteristics: mass (M), electric charges (Q) and angular momentum (J). Roughly speaking, the reason that these properties are observed is that they are associated with long-range fields which can exert influence at large distances. No other properties of the matter which went into forming the back hole are preserved. This aspect of black holes is expressed by John A. Wheeler in these oft-quoted words: "A black hole has no hair". Although black holes are mathematically very complex, they are structurally simple.

Historically, soon after Schwarzschild obtained the spacetime geometry outside a spherical object of mass M, H. Reissner in 1916 and G. Nordstrom in 1918 independently solved Einstein's equations and found the spacetime geometry outside a spherical object of mass M and charge Q. Then in 1963, after a gap of 45 years, Roy P. Kerr found a solution for a black hole with mass M and angular momentum J. Some two years later E. T. Newman and others obtained solutions involving M, J, and Q, all the possible characteristics that could be possessed by black holes.

The Kerr-Newman spacetime has the line element

$$d\tau^2 = \frac{\Delta}{\rho^2}\left[dt - \frac{1}{c}a\sin^2\theta d\varphi\right]^2 - \frac{\sin^2\theta}{\rho^2}\left[\frac{1}{c^2}(r^2 + a^2)d\varphi - adt\right]^2 - \frac{\rho^2}{c^2\Delta^2}dr^2 - \frac{1}{c^2}\rho^2 d\theta^2 \tag{27}$$

where

$$\Delta \equiv r^2 - \frac{2GM}{c^2}r + a^2 + \frac{GQ^2}{c^4} \tag{27a}$$

$$\rho \equiv r^2 + a^2\cos^2\theta; \qquad a \equiv \frac{J}{Mc}\,. \tag{27b}$$

Kerr-Newman's solution has rotational symmetry about the axis $\theta = 0$; none of the metric coefficients depends on the cyclic coordinate φ. It is, moreover, stationary: none of the metric coefficients depends on the coordinate t which is time for an observer at infinity. For $a = 0$ and $Q = 0$, Kerr-Newman's solution reduces to Schwarzschild's solution:

$$d\tau^2 = \left(1 - \frac{2GM}{c^2r}\right)dt^2 - \frac{1}{c^2}\left[\left(1 - \frac{2GM}{c^2r}\right)^{-1}dr^2 + r^2d\theta^2 + r^2\sin^2\theta d\varphi^2\right]. \tag{28}$$

For J = 0, but M = 0 and Q = 0, the Kerr-Newman solution reduces to the Reissner-Nordstrom solution

$$d\tau^2 = \left(1 - \frac{2GM}{c^2 r} + \frac{GQ^2}{c^4 r^2}\right)dt - \frac{1}{c^2}\left[\left(1 - \frac{2GM}{c^2 r} + \frac{GQ^2}{C^4 r^2}\right)^{-1} dr^2 + r^2 d\theta^2 + r^2 \sin^2\theta d\varphi^2\right]. \quad (29)$$

When Q = 0, the Kerr-Newman solution reduces to Kerr's solution:

$$d\tau^2 = 1 - \frac{2GM}{c^2\rho^2}dt^2 - \frac{2}{c}\left(\frac{2GMr}{c^2\rho^2}\right)a \sin^2\theta dt d\varphi - \frac{1}{c^2}\frac{\rho^2}{\Delta}{}' dr^2 - \frac{1}{c^2}d\theta^2 - \frac{\Lambda}{c^2\rho^2}\sin^2\theta d\varphi^2 \quad (30)$$

where

$$\Delta' = r^2 + a^2 - \frac{2GM}{c^2}; \qquad \Lambda = (r^2 + a^2)^2 - a^2\sin^2\theta . \quad (30a)$$

The Kerr-Newman metric, like Schwarzschild's, has an event horizon which is spherical in shape, and its surface area A is given by the formula

$$A = 4\pi\left(r_+^2 + a^2\right) \quad (31)$$

whereas its "radius" is given by

$$r_+ = \frac{1}{2}\left[r_g + \sqrt{r_g^2 + 4a^2 - 4q^2}\right] \quad (32)$$

where $q = G^{1/2}Q/c^2$. Notice that the area A differs from the Euclidean formula, and this is due to the fact that the geometry of the black hole is non-euclidean. For r_+ to be a real number, the quantity under the square root must be positive; that is, $r_g^2 + 4a^2 - 4q^2 > 0$. If this quantity is positive, there appears to be another horizon at

$$r_- = \frac{1}{2}\left[r_g - \sqrt{r_g^2 + 4a^2 - 4q^2}\right]. \quad (33)$$

However, since $r_- < r_+$, the outside observer is concerned only with r_+.

If the quantity under the square root is negative, there will be no event horizon and we shall have a "naked singularity", i.e. a singularity that will be visible and communicate to the outside world. Do black holes of this type exist? Or is there a cosmic censorship that prevents naked singularity to happen? This is yet an open question.

There is another surface of physical significance surrounding the event horizon. This surface is known as the static limit and it is not spherical. It is bun-shaped, being flattened at the poles which lie on the axis of rotation of the black hole. Figure 3 shows the equatorial and meridional sections of the black hole. These show that the surface of the static limit touches the event horizon at the poles. The space between the event horizon and the static limit is called the ergosphere. At a latitude α, the radial coordinate for the ergosphere is given by

$$r_\alpha = \frac{1}{2}\left[r_g + \sqrt{r_g^2 - 4q^2 - 4a^2\sin^2\alpha}\,\right] \tag{34}$$

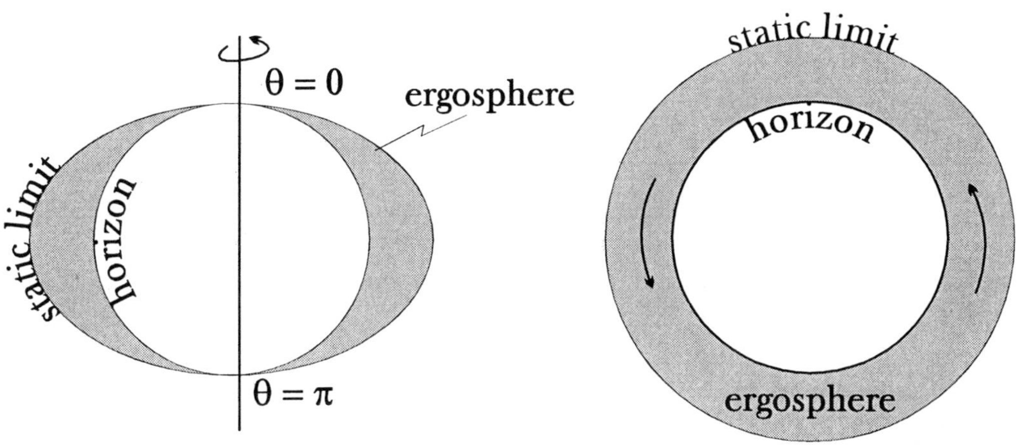

(a) The meridional section   (b) The equatorial section

Figure 3: A rotating black hole.

Within the static limit nothing can stand still, because the space-time around a rotating object is dragged along with it. (The effect is known as the Lense-Thirring dragging of inertial frames). To understand this, consider the atmosphere dragged along with the earth's rotation. Not only bodies on the earth rotate with earth but also bodies in the air. Birds flying up in the air and down again return to their starting places: they do not find earth's surface to have shifted eastward while they were in the midair. That is because the atmosphere (in which the bird flies) is carried along with the earth and therefore the bird does not find any relative displacement with respect to the earth. In a manner somewhat analogous to this example, the spacetime around a rotating object is carried along with it. To test this effect on the space-time around the earth an experiment has been proposed which involves launching a gyroscope in orbit round the earth. Normally the axis of a gyroscope is fixed in space. However, if the above-mentioned general relativistic effect is present, the axis should precess about a fixed direction. The predicted precession rate is about 7" per year at a height of 800 km, and this is not too small to be measured.

Material particles and photons can cross the static limit in either direction. Hence, unlike the event horizon, the static limit does not prevent outward leakage of information.

The occurrence of the two separate surfaces (the event horizon and the static limit) in the Kerr-Newman geometry allows energy extraction from a black hole. This possibility derives from the fact that in the ergosphere (the space between the event horizon and the static limit) the coordinate t, which is time-like external to the static limit, become space-like, and so the components of the four-momentum in the t-direction, which is the conserved energy for an observer at infinity, becomes space-like in the ergosphere; they can accordingly assume here negative values. These circumstances give rise to an unexpected energy extraction possibility. Roger Penrose in 1969 outlined a thought experiment to demonstrate this. As shown in Figure 4, the process involves dropping an element of matter into the ergosphere (where it acquires energy and angular momentum from the black hole), arranging for it to split (in the ergosphere) into two parts in such a way that one part, as judged by an observer at infinity, has a negative energy. Conservation of energy requires that the other part acquires an energy that is in excess of that of the origin element. If the part with the excess energy escapes along a geodesic while the other part crosses the event horizon and falls into the black hole, then we should have extracted some rotational energy of the black hole by reducing its angular momentum.

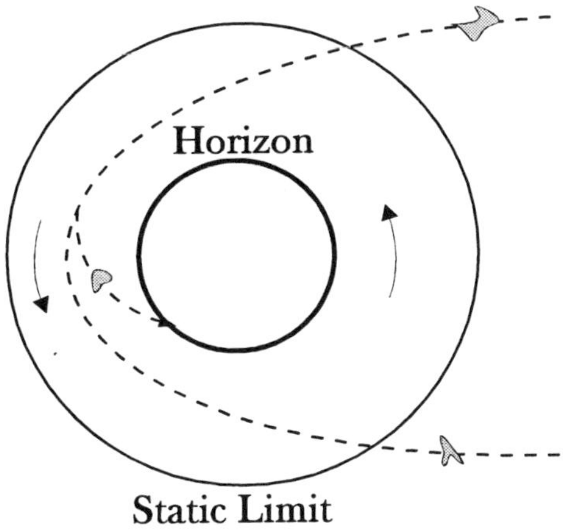

Figure 4: The Penrose mechanism.

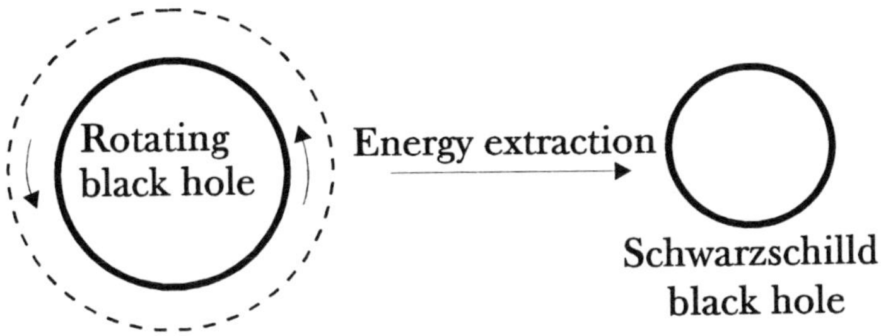

Figure 5: The Schwarzschild black hole represents the final, irreducible state.

The Penrose mechanism can continue until the black hole has given away all its rotational energy. The ergosphere then no longer exists, and we arrive at the non-rotating Schwarzschild black hole. Thus the Schwarzschild black hole represents the final, irreducible state in which external processes can only increase the energy of the black hole instead of decreasing it (Figure 5).

In considering the energy that could be released by interaction with black holes, Steven Hawking discovered an important theorem in 1971. This, **the area theorem**, states that *in interactions involving black holes, the total surface area of the event horizon of a black hole can never decrease (in the absence of quantum effects); it can at best, remain* *unchanged (if the conditions are stationary).* To illustrate the area theorem, Hawking considered the following example. Imagine two spherical (Schwarzschild) black holes, each of mass M/2, coalescing to form a single spherical black hole of mass M′. The surface area of the two black holes before merger is

$$dA_1 = 2 \times (4\pi r_g{}^2) = 8\pi[2G(M/2)/c^2]^2 = 8\pi(GM/c^2)^2 .$$

In the final state, the area of the Schwarzschild black hole is

$$dA_2 = 4\pi(2GM'/c^2)^2 = 16\pi G^2M'^2/c^4 .$$

The area theorem requires that $dA_2 > dA_1$:

$$16\pi(GM'/c^2)^2 > 8\pi(GM/c^2)^2 ,$$

from which it follows that

$$M' > M/\sqrt{2} .$$

Hence the maximum amount of energy that can be released in such a coalescence is

$$(1 - 1/\sqrt{2})Mc^2 = 0.293\ Mc^2 .$$

In practice, the actual amount may be much less; but it is clear that the processes of the type considered have the potentiality for releasing a far larger fraction of the rest mass as energy than nuclear processes.

The Kerr-Newman black hole appears to be the most general type of black hole. But rotating black holes are most likely Kerr black holes. It is unlikely that black holes can accumulate significant electrical charge, at least for long. If a black hole were formed with, say, a strong net negative charge, it would quickly attract positive charges in its vicinity and repel negative ones. Over a period of time the original negative charge would be neutralized. Thus, the only practical observables of a black hole are its mass and angular momentum. That is, a rotating black hole is most likely a Kerr black hole, and we should set $Q = 0$ in the above discussion.

8-6 Thermodynamics of black holes

Hawking's area theorem opened up a new avenue in the study of black hole physics. Regrading the surface area of the event horizon of a black hole, its behavior is analogous to the behavior of a quantity known as entropy in thermodynamics, a science of the behavior of energy and information in physical systems.

The area theorem is very similar, in wording, to the second law of thermodynamics, which states that the entropy of a closed system cannot decrease; in the course of any process it must either increase or remain unchanged. By entropy we mean the "unavailability" of energy — energy is not available in a suitable form for useful work; alternatively, we can regard the entropy of a system as being a measure of the disorder of that system, or of lack of information of its precise state. As entropy increases, the amount of energy available for useful work decreases, or the amount of information about the state of a system decreases.

The analogy or similarity between the behavior of entropy and the properties of event horizons led Jacob Beckenstein in 1972 to speculate that the analogy might provide a meaningful link between black-hole physics (gravitation) and thermodynamics, two apparently disparate sciences. Could a black hole possess entropy? For this idea to have merit, however, it must be possible: (1) to define precisely what is meant by the "entropy" of a black hole; and (2) to associate the concept of temperature with a black hole.

Beckenstein proposed to define the entropy of the black hole on the basis of the so-called "no-hair theorem" which was proved by Carter, Hawking, Israel, Robinson, and Price. A black hole has only three distinguishing features: mass, angular momentum and electric charge. Beckenstein proposed and argued that the entropy of a black hole could be described in terms of the number of possible internal states which correspond to the same external appearance. In other words, the more massive the black hole, the greater the number of possible configurations which went into its formation, and the greater the loss of information. A vast amount of information is lost in the formation of a black hole. The area of the event horizon is proportional to the square of the mass of the black hole: the more massive the black hole the greater its event horizon, and the more massive the black hole the larger the area of its event horizon, so it seems reasonable to regard the entropy of a black hole as being proportional to the area of its event horizon.

To associate a temperature with a black hole, we can use the analogy of thermodynamics again. The temperature of a body is uniform at thermodynamic equilibrium (this is often called the zeroth law of thermodynamics). In black hole physics, a corresponding state exists for axisymmetric stationary black holes. (Stationarity means a state does not change with time, and axisymmetry implies symmetry about some axis, e.g. the axis of rotation.) For such a black hole, the surface gravity is the same all over the event horizon. This analogy was developed further in 1973 by J. Bardeen, B. Carter and S. Hawking, who showed that the surface gravity of a black hole played a role analogous to the concept of temperature in thermodynamics. The surface gravity at the event horizon of a black hole is inversely proportional to its mass and, if the analogy is carried through, this implies also that the temperature of a black hole is inversely proportional to its mass — the less massive the black hole, the "hotter" it would be. This identification is reinforced by the following consideration. For a Kerr-Newman black hole, its "radius" and the area of its event horizon are given by

$$r_+ = \frac{1}{2}\left[r_g + \sqrt{r_g^2 - 4a^2 - 4q^2}\right] \tag{35}$$

$$A = 4\pi\left(a^2 + r_+^2\right) = 2\pi\left[r_g^2 + r_g\sqrt{r_g^2 - 4q^2 - 4a^2}\right] - 4\pi q^2 \tag{36}$$

respectively, where

$$a = J/Mc; \qquad q = \frac{\sqrt{G}Q}{c^2}. \tag{36a}$$

Suppose we make small changes in the mass (M), angular momentum (J), and the electric charge (Q). This will result in a change of its surface area of the event horizon also. Simple calculation gives the following differential relation which relates these changes:

$$\delta(Mc^2) = \frac{\kappa c^2}{8\pi G}\delta A + \frac{AC^2}{a^2 + r_+^2}\delta J + \frac{r_+ Qc^2}{a^2 + r_+^2}\delta Q \tag{37}$$

where κ is the surface gravity

$$\kappa = \sqrt{r_g^2 - 4q^2 - 4a^2}/2(a^2 + r_+^2) . \tag{37a}$$

We now compare the differential relation (37) with the thermodynamic relation

$$\delta U = T\delta S - p\delta V \tag{38}$$

where T is the temperature, S the entropy, U the internal energy of the system, and p and V are the pressure and volume of the system. This thermodynamic relation relates the increase in the internal energy of the system to the change in entropy and to the work done by (or against) the pressure. For example, if pressure puts in work and compresses the system so that the change in the volume δV is negative, this work increases the internal energy of the system. If we read these two relations in conjunction with the Second Law of thermodynamics and the area theorem of a black hole

$$\delta S \geq 0, \qquad \delta A \geq 0 \tag{39}$$

the analogy becomes clear. The area A is analogous to entropy S, the surface gravity κ is analogous to the temperature T, and the work done in changing the angular momentum or the electric charge of the black hole is analogous to the work done in changing the volume of the thermodynamic system. In each case the net result of the two relations is to change the energy of the system: the black hole or the thermodynamic system.

The above discussion suggests that the temperature of a black hole with a finite mass is nonzero. But how can a black hole have a finite temperature? Thermal bodies with a finite temperature should emit thermal radiation in accordance with Planck's law. Yet according to the classical definition of a black hole, matter and energy could only fall into black holes; nothing could emerge from them. S. Hawking realized that this classical conclusion might not hold quantum mechanically. Therefore he investigated the quantum behavior of matter in the neighborhood of a black hole, and in 1974 he found a way out of the paradox. He discovered that black holes would appear to emit particles such as photons, electrons and neutrinos, and that to a distant observer this radiation would have a thermal spectrum, the same kind of spectrum emitted by a black body. This startling discovery opened up the way for the establishment of links between gravitation, thermodynamics and quantum theory.

The Hawking effect involves a quantum concept of vacuum. In classical physics, a vacuum implies absence of everything. But the quantum vacuum is a swarm of particles and antiparticles which are constantly being created and destroyed. These particles and anti-particles are considered to be virtual in the sense that they don't last long enough to be observed. This quantum concept of vacuum is related to Heisenberg's uncertainty principle. Due to the dual nature (wave and particle) of subatomic particles, a certain amount of uncertainty enters into the description of these particles. We cannot determine simultaneously, for example, the precise position and momentum of a subatomic particle; we can determine only the probabilities of finding particles in particular places and having particular momenta. Similarly, we cannot know precisely the exact energy of a quantum system at every moment in time. Over short time intervals, there can be great uncertainty about the amount of energy in the subatomic world. Specifically, if ΔE is the uncertainty in energy measured over a short time interval Δt, then

$$\Delta E \times \Delta t \geq \frac{h}{2\pi} \tag{40}$$

where h is Planck's constant. We now combine this with Einstein's equation $E = mc^2$. There is nothing uncertain about c, the speed of light. Therefore any uncertainty in the energy of a physical system can be attributed to an uncertainty Δm in the mass. Thus

$$\Delta E = c^2 \Delta m \ . \tag{41}$$

Combining these two expressions, we obtain

$$\Delta m \times \Delta t \geq \frac{h}{2\pi c^2} \ . \tag{42}$$

This result is astonishing. It means that, in a very brief interval Δt of time, we cannot be sure how much matter there is in a particular location, even in vacuum. A quantum-mechanical vacuum, in fact, is a very busy place. At any place it is possible to spontaneously create a particle-antiparticle pair. This requires an energy equal to $2mc^2$, where m is the mass of the particle or antiparticle. This pair can only exist for at most a time $h/\Delta mc^2$. Before that time is up they must find each other and annihilate (Figure 6a). We can therefore think of the quantum vacuum as being made up of continuously appearing and disappearing particle-antiparticle pairs. The importance of vacuum fluctuations in electromagnetic processes has long been experimentally confirmed. If electron-positron pairs are created near a real electron, the electron will attract virtual positrons and repel virtual electrons. The resulting cloud of excess positive charge surrounding the real electron cancels most of its bare charge, leaving the small net charge, -e, that is measured by experiments carried out at large distances from the electron. Sampled at closer distances where the layer of shielding is partially penetrated, the measured charge would increase in magnitude. Precisely such an effect has been detected in the so-called Lamb shift of the spectral lines of the hydrogen atom.

Hawking investigated how these vacuum fluctuations interact with a black hole (Figure 6b). The pair of thick lines are the boundaries of the black hole. The gravity of the black hole acts on the virtual pairs and it could produce the different types of effects. Both members of the pair may fall into the black hole, or one member may fall and the other escape, or the pair annihilated as it would be in the absence of the black hole. It is the second effect which is of interest. Receiving the escaped member by a distant observer gives him the impression that the black hole emits this particle.

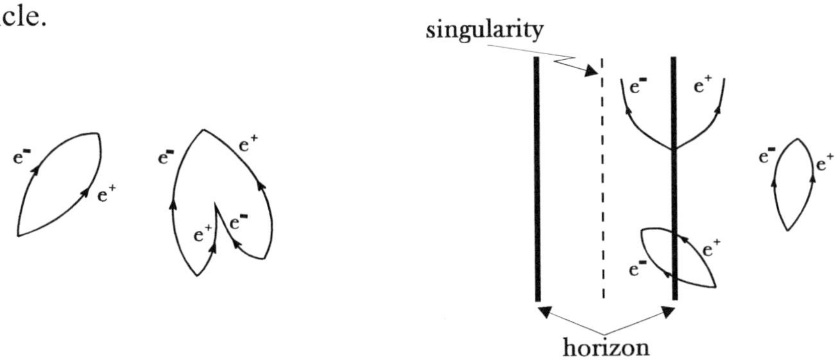

(a) Virtual pairs

(b) Pair production near a black hole

Figure 6: Quantum vacuum fluctuations.

Quantum mechanical calculation can tell the relative chances of occurrence of these various possibilities. Hawking performed such a calculation, and he discovered that the statistical effect of many such emissions leads to the emergence of particles with a thermal spectrum:

$$N(E) = \frac{8\pi}{c^3h^3}\frac{E^2}{e^{4\pi^2E/\kappa h} - 1} \tag{43}$$

where N(E) denotes the number of particles of energy E per unit energy band per unit volume. Notice the similarity between this formula and the Planckian distribution for thermal radiation from a black body. It was this similarity which led Hawking to conclude that a black hole radiates as a black body at the temperature T

$$T = \frac{h}{4\pi^2 kc}\kappa \tag{44}$$

where k is the Boltzmann constant. Note that the physical behavior of surface gravity and temperature is not merely analogous, it is identical. A similar relationship exists between the area of the event horizon A of the black hole and its entropy S

$$S = (\pi k/2\pi G)A\ . \tag{45}$$

For a Schwarzschild black hole, the surface gravity is

$$\kappa = GM/r_g^2 = c^4/4GM \tag{46}$$

and its temperature T can be expressed in terms of mass M

$$T = \frac{hc^3}{8\pi kGM} = 6 \times 10^{-8}\frac{M_\odot}{M}\,K\ . \tag{47}$$

To estimate the power radiated, we assume that the surface of the black hole radiates according to the Stefan-Boltzmann law

$$\frac{dP}{dA} = \sigma T^4 = \frac{\pi^2}{60}\frac{k^4}{h^3c^2}T^4\ . \tag{48}$$

Multiplying by the area A $(= 4\pi r_g = 16\pi G^2M^2/c^2)$, substituting the temperature Eq.(44), and noting that the power radiated corresponds to a decrease of mass M by $P = -(dM/dt)c^2$, we obtain the differential equation for the mass of a radiating black hole as a function of t:

$$\frac{dM}{dt} = -\frac{\pi^3}{15360}\frac{hc^4}{G^2M^2} . \qquad (49)$$

Using the equation $Pt = Mc^2$, we can estimate the duration of the radiation:

$$t \simeq Mc^2/P \simeq G^2M^3/hc^4 \simeq 10^{-20}M^3 \text{ sec} \qquad (50)$$

where M is in kg.

These results indicate that the Hawking effect for black holes with several solar masses leads to a negligible temperature and power output. The more massive the black hole, the slower the rate at which mass is being lost. A black hole with a mass comparable to that of the sun would have a temperature of about 10^{-7} K and would emit radiation at the totally negligible rate ~ 10^{-16} erg/s.

But affairs are different if the black hole has a much smaller mass. A mini black hole of the size of a proton would contain 10^{12} kg mass and have a temperature of about 10^{11} K. It would be emitting electrons, positrons, photons, neutrinos and other kinds of particles with a power output of some 6,000 megawatts. As a black hole loses mass, its temperature increases. The hotter it becomes the faster it radiates, and the faster it radiates the faster it loses mass. As the mass of the black hole becomes very small the process escalates very rapidly until, in the end, the black hole radiates away the last of its mass-energy in a catastrophic explosion. At our present state of knowledge we cannot predict precisely what would occur in the final stages of explosion; but it is certain that the final explosion would result in the release of a tremendous burst of high-energy gamma rays. What would be left behind after the explosion? We do not have a theory capable of explaining what happens when a black hole shrinks within the Planck radius (10^{-35} m), so this question lies in the area of speculation.

A mini black hole of about 10^{12} kg would be expected to decay over a period of about 10^{10} years, in the same order as the estimated age of the Universe. Thus, we expect some such mini black holes might be exploding now, resulting in the release of tremendous bursts of high-energy gamma rays. The detection of an exploding black hole would be a discovery of utmost importance: (1) It demonstrates the validity of the Hawking theory and of the links between gravitation, thermodynamics and quantum theory. (2) Because different theories of particle physics make quite different predictions about the properties of such an explosion, analysis of the energy emissions would provide crucial information about the nature of fundamental particles. So far no such explosions have been detected. The primordial mini black holes still remain a theoretical possibility.

8-7 Observing black holes

We cannot observe black holes directly; we can only detect them by their interaction with other material. Stellar-mass black holes might be detected if they are members of binary systems.

(1) Searches for invisible black holes in binary (or multiple) stellar systems.

A binary system consisting of a normal star and a black hole, circling each other at a large distance (large, compared with the star's diameter) may be detected through the Doppler shift of the star's spectral lines. Aquarii is a suspected candidate. But the problem happens to be just as difficult as it is uncertain. An "invisible" massive component need not necessarily be a black hole; it might possibly be embedded in a dust cloud, making it invisible.

(2) Searches for black holes that are powerful sources of x-rays in binary systems.

A close binary system consisting of a black hole and a normal star gives rise to a new phenomena. The visible component could fill its Roche lobe, and a powerful stream of gas would fall into the black hole. Since the gas stream would carry much angular momentum along with it, the gas would form a rapidly spinning disk around the black hole. The particles comprising the accretion disk would orbit the black hole approximately according to Kepler's law. Viscosity would cause the particles of the disk to lose angular momentum continuously, and some of them would gradually settle into the black hole. As the gas sinks into the black hole the temperature of the inner zone of the accretion disk may reach several million degrees. Such a disk could be a strong source of x-rays. To a first approximation, the power and spectrum of the radiation would be the same as for neutron stars that are x-ray pulsars. A decisive test for distinguishing a black hole from a neutron star is the determination of the mass of the x-ray source. Binary x-ray sources are the best suspects for containing a black hole, and, for example, the bright x-ray source Cygnus X-1 has been considered by many as a possible black hole. Many investigators believe that the compact x-ray component of the Cygnus X-1 system has a mass in excess of 6 $M_{\odot}$, and accordingly should be a black hole.

Observations have also led to the belief that supermassive black holes might be lurking in the centers of globular clusters and galaxies. The clues were found several years ago in star counts. A count of stars as a function of distance from the center of the cluster showed an unexpectedly high concentration of stars close to the center in three different globular clusters. This concentration could be explained by the presence of a black hole of several thousand solar masses in the center of the cluster. Similarly, a photometric study of the elliptic galaxy M87 showed the distribution of its stars does not agree with the normal distribution found in other galaxies, and a steep increase in the density of stars was found in a 4 light-year neighborhood around the galactic center. Again, the observational data of M87 could be explained by the presence of a massive black hole of about 5 billion solar masses in the center of M87. Current theory cannot explain the formation of such a massive black hole in the center of a galaxy. Thus, confirmation of the existence of such a massive black hole would lead to fundamental changes in astronomy.

Problems

(8-1) Calculate the density that different masses (proton, earth, Sun, Milky Way galaxy) reach at their Schwarzschild radii. What conclusion can be drawn from this data?

(8-2) Find the radii of circular orbits for a particle in the field of a Schwarzschild black hole. Show that the radius of the stable orbit closest to the center is given by

$$r = 3r_g \ .$$

(8-3) Show that the Kruskal transformation (also known as Kruskal-Szekeres transformation) given by Eqs.(23) and (24) converts the Schwarzschild metric to the form given by Eq.(25).

(8-4) Show that once a rocket ship crosses the gravitational radius (horizon) of a Schwarzschild black hole, it will reach $r = 0$ in a proper time $\tau \leq \pi M$, no matter how the engines are fired.

(8-5) Show that the surface area of the horizon of a Kerr-Newman black hole is

$$4\pi\left(\left[M + (M^2 - Q^2 - a^2)^{\frac{1}{2}}\right]^2 + a^2\right) .$$

(8-6) Show that Kepler's law

$$\Omega^2 = M/r^3$$

holds for circular orbits around a Schwarzschild black hole, if r is the curvature coordinate radius, and Ω is the angular frequency as measured from infinity. Derive an analogous law for equatorial orbits around a Kerr black hole of specific angular momentum a.

Chapter 9

Elements of Cosmology

9-1 Introduction

Einstein's general relativity is a satisfactory theory of gravitation and it provides a space-time structure whenever the matter distribution is given. Thus, if the average distribution of matter in the universe is put into Einstein's field equations, the average space-time structure of the whole universe may be deduced. This is a very interesting exercise and it is part of the subject of cosmology.

In general, cosmology is the study of the dynamical structure of the universe and seeks to answer questions regarding the origin, the evolution, and the future behavior of the universe as a whole. Historically, after establishing his general theory of relativity in 1916, Einstein promptly applied his theory to problems in cosmology, and published a first paper on relativistic cosmology in 1917. At that time, cosmology was the only field in which the significance of the general relativity could be fully manifested.

In this chapter and in the next, we will be studying cosmology. Since we are considering phenomena on the cosmological scale, we have to make assumptions about which physical processes are dominant on this scale and we need to develop models for the universe. These have to be rooted both in physical theory — in this case the general theory of relativity and the laws of physics known today — and in observations. The two basic phenomena that allow us to probe the large-scale structure of the universe are the following:

(1) the expansion of the universe (Hubble's law);

(2) the cosmic microwave background radiation.

9-2 Basic observations

The universe was for a long time considered static and homogeneous. It is easy to see that a finite static universe is not stable: as Newton wrote in 1662 in a letter to R. Bentley, "If the matter of our sun and planets and all the matter of the universe were evenly scattered throughout all the heavens, and every particle had an innate gravity toward all the rest, and the whole space throughout which this matter was scattered was but finite, the matter on the outside of this space would, by its gravity, tend towards all the matters on the inside and, by consequence, fall down into the middle of the whole space and there compose one great spherical mass." On the other hand, in a static universe infinitely extended, we meet the difficulty known as "Olbers' paradox". The paradox arises as follows. Consider an infinite, static, universe uniformly filled with stars of luminosity L and of space density n. The apparent intensity (or flux density) of a star at the earth is given by the inverse square law $L/4\pi r^2$. The number of stars in the distance range r to r + dr is $4\pi r^2 n dr$, and consequently the total intensity of radiation arriving at the earth is

$$I = \int_0^\infty \frac{L}{4\pi r^2} 4\pi r^2 n \, dr = Ln \int_0^\infty dr \rightarrow \infty$$

i.e., in a uniform, infinite, static universe, the total brightness of the sky would be infinite! This deduction is not quite right since the light of distant stars will be obscured by closer stars. The correct conclusion is that a uniform, infinite, static universe must be in thermal equilibrium, i.e., all bodies must eventually arrive at the same temperature. Thus, the earth would be heated by incoming radiation to several thousand degrees and would in turn radiate just as much as it receives. The sky would not know the difference between day and night.

(A) Hubble's law and the expansion of the universe

The solution to Olbers' paradox was not realized until the expansion of the universe had already been discovered. By 1925 V. M. Slipher had shown that the spectral lines of the distant galaxies are shifted toward the red end of the spectrum, and interpreting this as a Doppler shift, he concluded that the galaxies are receding from us. The red shift has a doubly weakening effect. In the first place, as the wavelength of the incoming waves is increased, their frequency is reduced; this diminishes their energy according to Planck's formula: energy = Planck constant x frequency. Second, the lowering of the frequency means that fewer arrive in one second, so that the energy received is still further reduced. Furthermore, expansion implies a finite age for the universe. We can see stars that are close enough for their light to have reached us in this (finite) time. Thus, there is a finite cutoff to the number of stars that can contribute to the sky brightness, and the problem is solved. (E. Harrison has recently pointed out that finite luminous lifetime of the stars is more significant than the expansion of the universe.)

In 1929 Hubble showed that the velocity v of the galaxies, as measured by the redshift of their spectral lines, increases linearly with distance r. Hubble's work has been strongly confirmed by subsequent work. This redshift-distance relation is known as the law of redshift or Hubble's law:

$$\vec{v} = H\vec{r}$$

where H is a constant (Hubble constant): H = (55 ± 7) km/s/Mpc. The word <u>constant</u> means of course that H does not depend on the magnitude or on the direction of the vector $\vec{r}$. H may depend however on time.

Hubble's law asserts that the universe is in uniform expansion. By definition, in uniform expansion $\Delta r/r$ = constant in any given time interval Δt, i.e. $\Delta r/\Delta t \propto r$, or $v \propto r$. Since the universe is in uniform expansion, there are no privileged positions, and so an observer moving together with any galaxy sees the surrounding galaxies receding from him. As shown in Figure 1, consider the motion of galaxy G as seen by 0 and G′. As seen by 0,

$$\vec{v} = H\vec{r}, \quad \vec{v}' = H\vec{r}'$$

So $\vec{v}-\vec{v}'$, the velocity of G relative to G′, can be expressed as

$$\vec{v} - \vec{v}' = H\vec{r} - H\vec{r}' = H(\vec{r} - \vec{r}')$$

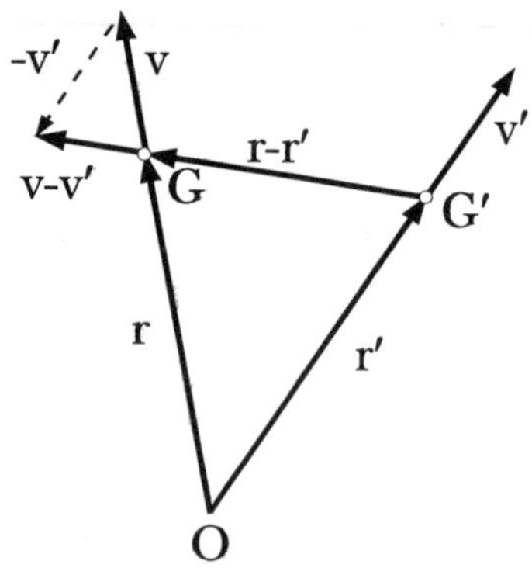

Figure 1: Expansion as seen by O and G'.

i.e., G' also sees G, and therefore all the other galaxies, receding from itself. Although we have used Euclidean geometry, and ignored possible changes in H with time, the general result nonetheless holds: each galaxy sees all the others receding from itself.

In conclusion of this section, we want to mention another aspect of Hubble's law. Since galaxies recede from us with a velocity directly proportional to their distance, there must be a point at which the velocity reaches and even exceeds the speed of light. The set of all those points is called the world horizon, which is about

$$r_{wh} = c/H = 2 \times 10^{10} \text{ light years}$$

where H was set equal to 55 km/sec/Mpc. Galaxies that are located at or beyond the world horizon are invisible to us.

(B) The microwave background radiation & the isotropy of the universe

The microwave background radiation was discovered by A. Penzias and R. Wilson in 1965. It was soon established that this radiation is extremely isotropic: it has the same intensity in all direction to a precision of better than one part in a thousand on all scales from about 1 arcmin to 360°. No matter what the origin of the radiation, this is compelling evidence that our starting point should be the study of isotropic models of the universe.

The microwave background radiation has a black body distribution to a high degree of precision with a radiation temperature of about 2.9 K (Figure 2). This implies that at some stage the radiation must have been in equilibrium with matter at a particular temperature.

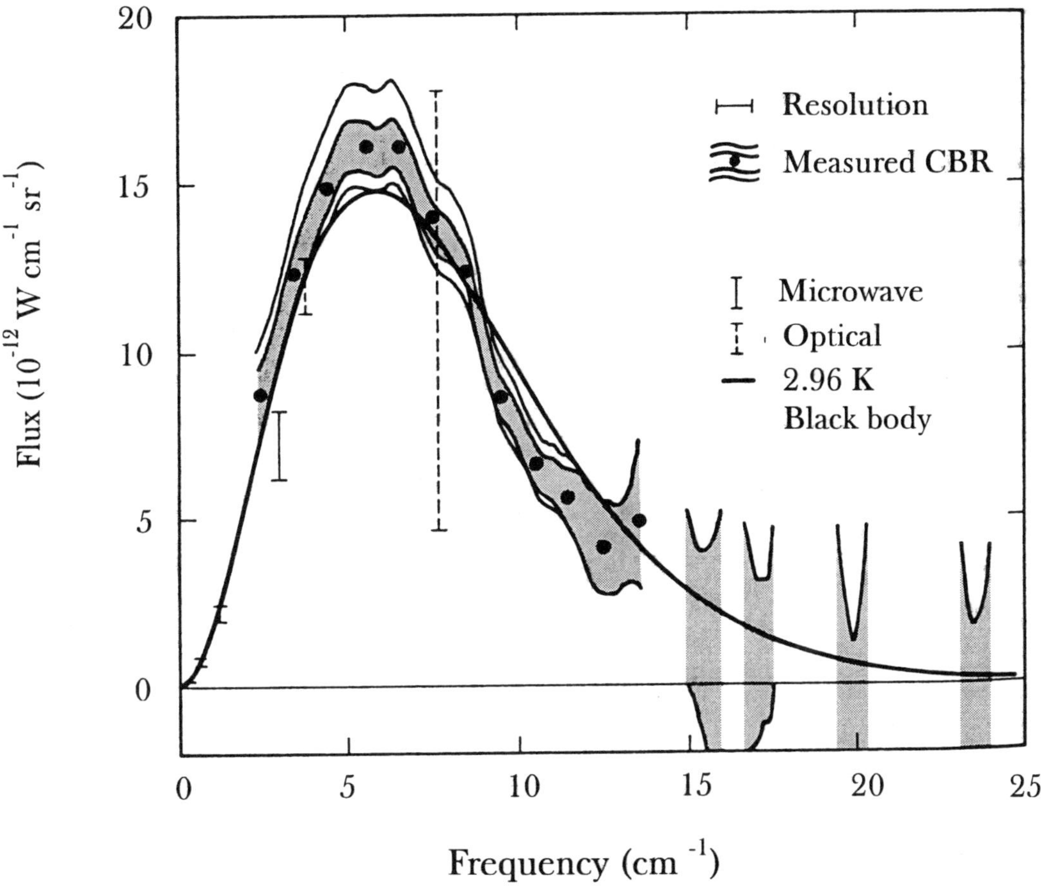

Figure 2: The spectrum of the microwave background radiation.

Another feature of the microwave background radiation is that the radiation is rather massive. The inertial mass density of the radiation at 2.9 K is a T^4/c^2 g cm^{-3} = 6×10^{-34} g cm^{-3}, where $a = 4\sigma/c$, σ being the Stefan-Boltzmann constant. This is by far the most massive component of diffuse radiation in the universe. The average density of ordinary matter in the universe is about 10^{-30} to 10^{-29} g/cm^3 and therefore there is much more inertial mass in the matter than in the radiation at the present time. But we will see that there is good reason to believe that the universe was radiation-dominant in the past, and that the observed microwave background radiation is a fossil remnant of these ancient times.

9-3 The Robertson-Walker metric

(A) The cosmic fluid and fundamental observers

The data of the above section indicates that *the large-scale universe is both homogeneous and isotropic*. The word "large-scale" refers to the scale of many superclusters of galaxies. We shall adopt this as an assumption and it is often called the cosmological principle. Thus every point in the universe is equivalent to every other point; there are no preferred positions or directions in the cosmos.

The actual universe is clearly very non-homogeneous on the small scale and only rough limits can be put on its homogeneity on the large scale. Matter is concentrated in stars, which in turn are collected into galaxies or clusters of galaxies. We make no attempt to incorporate individual galaxies or clusters of galaxies into our description of the universe as a whole. Instead, we imagine matter in the universe as being smeared out into an idealized, smooth fluid, which we call the "cosmic fluid", devoid of shear-viscous, bulk-viscous and heat-conductive properties. This is an idealization and a good description so long as we take a large-scale view of the universe. In physics we encounter a similar situation when we study properties of a gas. To describe the properties of a gas we do not need to study the behavior of individual atoms and molecules. Instead we define various macroscopic quantities — density, pressure, temperature — and study the relations between them.

We call an observer who is at rest with respect to this cosmic fluid a fundamental observer. As the universe expands, the cosmic fluid shares in the expansion and the fundamental observer will be co-moving with the fluid. Every co-moving observer in the cosmic fluid sees the same picture of the universe i.e., in the co-moving frame of reference the universe looks isotropic and homogeneous. What is the form of the spacetime metric in the co-moving frame? It is obvious that we cannot use the static Schwarzschild metric for an expanding universe which is certainly not static. We now derive the form of the spacetime metric in the co-moving frame in the following section.

(B) The isotropic, homogeneous line element

For the time coordinate t we take the proper time measured by a fundamental observer co-moving with the fluid. As before, $x^{\circ} = ct$. The line-element must have the form

$$ds^2 = c^2dt^2 + 2g_{0i}c\,dt\,dx^i + g_{ij}dx^idx^j; \qquad i,j = 1,2,3\ . \tag{1}$$

Recall that the interval of proper time is the line element with vanishing spatial increments $dx^i = 0$. Eq.(1) clearly displays that t is the proper time. And from Eq.(1) we have

$$g_{00} = c^2\ . \tag{2}$$

Furthermore, a particle at rest in the co-moving frame (i.e., one with $u^i = 0$, i = 1, 2, 3) must satisfy the geodesic equation of motion. Recalling the equation of motion in the form given by Eq.(5-48), and using Eq.(2), we obtain:

$$\frac{d}{dt}u_\mu = 0\ . \tag{3}$$

And therefore

$$\frac{d}{dt}\left(g_{\nu\lambda}u^\lambda\right) = \frac{d}{dt}\left(g_{\nu 0}u^0\right) = \frac{d}{dt}g_{\nu 0} = 0 \tag{4}$$

which reduces, with the help of Eq.(2), to

$$\frac{d}{dt}g_{0i} = 0\ . \tag{5}$$

The validity of Eqs.(2) and (5) depend on, and only on, the choice of the co-moving frame.

We now apply the condition of isotropy to our line-element. The form of the line-element possessing central symmetry about one point has been already discussed in Chapter 6. Recalling this discussion and utilizing Eqs.(2) and (5) we can put ds^2 in the following isotropic form

$$ds^2 = c^2dt^2 - D(r,t)dr^2 - cE(r)dt\,dr - F(r,t)(d\theta^2 + \sin^2\theta\, d\varphi^2) \,. \tag{6}$$

We now try to eliminate the off-diagonal term E(r) from the line-element. To this end, we note that Eq.(2) remains unchanged under transformations

$$t' = t + f(x^i); \qquad x'^i = x^i \tag{7}$$

where $f(x^i)$ is an arbitrary function of spatial coordinates. We can check this easily by inserting Eq.(7) into Eq.(1). We now exploit this coordinate transformation to eliminate the term E(r). This is accomplished by choosing

$$f(x) = -\frac{1}{2}\int^r E(r)dr \quad \text{or} \quad dt' = dt - \frac{1}{2}E(r)dr \,. \tag{8}$$

The line element, dropping primes, is now brought into the form

$$ds^2 = c^2dt^2 - A(r,t)dr^2 - B(r,t)(d\theta^2 + \sin^2\theta\, d\varphi^2) \tag{9}$$

where $A(r,t) = -g_{11}$ and $B(r,t) = -g_{22}$ are unknown functions. The above line element possesses central symmetry about the origin $r = 0$. We now impose homogeneity, and consider infinitesimal spatial displacements

$$x^i \to x'^i = x^i + \epsilon^i(x), \qquad i = 1,2,3 \,. \tag{10}$$

ϵ is an infinitesimal parameter. Now homogeneity means that the line element retains its form under the transformation (10), i.e.,

$$g'_{ij}(x') = g_{ij}(x'), \qquad i,j = 1,2,3 \,. \tag{11}$$

Note that the results $g_{00} = c^2$ and $d(g_{0i})/dt = 0$ are unaffected by the transformation (10). We now apply the transformation (10) to the tensor transformation

$$g'_{\mu\nu} = \frac{\partial x^{\alpha}}{\partial x'^{\mu}} \frac{\partial x^{\beta}}{\partial x'^{\nu}} g_{\alpha\beta}$$

and obtain

$$g'_{ij}(x') = g_{ij}(x) - \epsilon \left(\frac{\partial \epsilon^{k}}{\partial x^{i}} g_{kj}(x) + \frac{\partial \epsilon^{k}}{\partial x^{j}} g_{ik}(x) \right) \tag{12}$$

and from Taylor series expansion we can further obtain

$$g_{ij}(x') = g_{ij}(x) + \varepsilon \xi^{k} \frac{\partial g_{ij}(x)}{\partial x^{k}} + O(\varepsilon^{2}) . \tag{13}$$

From Eqs.(11) and (13) we obtain

$$\frac{\partial \xi^{k}}{\partial x^{i}} g_{jk}(x) + \frac{\partial \xi^{k}}{\partial x^{j}} g_{ik}(x) + \xi^{k} \frac{\partial g_{ij}(x)}{\partial x^{k}} = 0 . \tag{14}$$

For $i = 1$ and $j = 1$ the above equation gives

$$2 \frac{\partial \xi^{1}}{\partial r} A(r,t) + \xi^{1} \frac{\partial A(r,t)}{\partial r} = 0 . \tag{15a}$$

Similarly for $i = 1$, $j = 2$ and $i = 2$, $j = 2$ we have

$$\frac{\partial \xi^{2}}{\partial r} B(r,t) + \frac{\partial \xi^{1}}{\partial \theta} A(r,t) = 0 \tag{15b}$$

and

$$2 \frac{\partial \xi^{2}}{\partial \theta} B(r,t) + \xi^{1} \frac{\partial B(r,t)}{\partial r} = 0 . \tag{15c}$$

Eq.(15a) requires that A'/A, where $A' = \partial A/\partial r$, be a function of r but not of time t. Hence A(r,t) must factorize as

$$A(r,t) = f(r)R^{2}(t). \tag{16}$$

The same argument when applied to Eqs.(15b) and (15c) leads to

$$B(r,t) = g(r)R^{2}(t). \tag{17}$$

So far, the r coordinate has not been completely fixed, since Eqs.(9),

(16) and (17) are unchanged under a transformation of the type $r' = F(r)$, with arbitrary $F(r)$. We are free to set

$$g(r) = r^2 \ . \tag{18}$$

This means that we have chosen the r coordinate such that the surface area of a sphere with radius r is equal to $4\pi r^2 R^2(t)$. When $R(t) = 1$ at a suitably chosen time, the surface area is $4\pi r^2$. The variables θ and φ have the usual meaning of angle variables in polar coordinates. Now from Eqs.(15a) and (16) we obtain

$$\xi^1 = C(\theta,\varphi) f^{-1/2}(r) \ . \tag{19}$$

Next eliminate ξ^2 from Eqs.(15b) and (15c) and use Eqs.(16) - (19) to obtain

$$\frac{1}{C}\frac{\partial^2 C}{\partial \theta^2} = \frac{r^2}{f^{1/2}}\frac{d}{dr}\left(\frac{1}{r f^{1/2}}\right) = D \tag{20}$$

where D is a constant. To determine D we note that in the neighborhood of $r = 0$ the transformation (10) should mimic the ordinary law for infinitesimal translation of the origin in flat space. If the translation is chosen to be along the polar axis (z-axis) this law is $\xi^1 \propto \cos\theta$. This gives $D = -1$. Then Eq.(20) gives the solution

$$f(r) = \frac{1}{1 - kr} \tag{21}$$

where k is a constant. Collecting the results (16) - (18) and (21) we can finally write the line-element (9) in the Robertson-Walker form:

$$ds^2 = c^2 dt^2 - R^2(t)\left[\frac{dr^2}{1 - kr^2} + r^2 d\theta^2 + r^2 \sin^2\theta \, d\varphi^2\right] \tag{22}$$

where $R(t)$ is a dimensionless scale factor (see below) depending only on time t, and k is the curvature constant, which describes the geometry of space at a particular instant of time. We will see in the following section that $k > 0$ corresponds to a space of positive curvature, $k = 0$ to normal flat space, and $k < 0$ to a space of negative curvature. k can be taken as +1, 0, or -1 by a suitable rescaling of the radial coordinate r. It is important to bear in mind that r is a co-moving coordinate, that is, it is a label for a particular galaxy and does not alter as the universe expands. Since the universe is isotropic, θ and φ also remain fixed for a particular galaxy. The time coordinate t measures the proper time for the galaxy (for a particular galaxy, $dr = d\theta = d\varphi = 0$, then $ds = cdt$). There is therefore a universal cosmic time t, which is the same for all observers at rest with respect to the local matter. In practice galaxies have random motions, so the local inertial frame has to be defined in terms of the average motion of galaxies over a sufficiently large region, or with respect to the microwave background radiation.

If we determine the proper distance ct between two galaxies from the equation $ds = 0$, we find

$$c\,dt = R(t)\,d\sigma$$

and so

$$c\,\Delta t = R(t)\,\Delta\sigma \tag{23}$$

where $d\sigma^2$ is the metric of the 3-dimensional space at a fixed value of the cosmic time t,

$$d\sigma^2 = \frac{dr^2}{1 - kr^2} + r^2(d\theta^2 + \sin^2\theta\,d\varphi^2)\ . \tag{24}$$

All measurements are understood to be made at the same epoch t. Now since r is a co-moving label and remains fixed for each galaxy, and θ and φ remain fixed for isotropic motion, then $d\sigma$ is fixed, and the proper distance between two galaxies is just scaled by the function R(t) as t varies. For this reason R(t) is called the scale-factor: R increases or decreases as the universe expands or contracts.

9-4 Properties of the Robertson-Walker metric

Now let us examine the nature of the geometry of the three-dimensional space at a fixed cosmic time t. From Eq.(22) the element of length dL is

$$dL = R^2(t)\,d\sigma^2 \equiv \gamma_{ij}dx^i dx^j \tag{25}$$

where $d\sigma$ is given by Eq.(24). The three-dimensional Riemann tensor corresponding to Eq.(25) is

$$R_{ijkl} = \frac{k}{R^2}(\gamma_{ik}\gamma_{jl} - \gamma_{jk}\gamma_{il}) \tag{26}$$

(calculated from Eq.(4-54) with indices restricted to run over 1, 2, 3), and the corresponding Ricci tensor and curvature scalar are

$$R_{ij} = \frac{2k}{R^2}\gamma_{ij} = \frac{1}{3}R^k_{\ k}\gamma_{ij}; \quad R^k_{\ k} = \frac{6k}{R^2}\ . \tag{27}$$

Thus the curvature properties of the space are as specified by the constant $k/R^2(t)$. If $k > 0$ we have a space of constant positive curvature; if $k < 0$, a space of constant negative curvature. If $k = 0$ we have the usual flat space.

We first consider geometry of a space with $k > 0$ (i.e., having a constant positive curvature). We can gain additional intuitive insight into the geometry with the following new variables:

$$x = Rr\sin\theta\cos\varphi; \quad y = Rr\sin\theta\sin\varphi; \quad z = Rr\cos\theta\ . \tag{28}$$

We also introduce a redundant, extra variable x_4 by the relation

$$x^2 + y^2 + z^2 + x_4^2 \equiv R^2/k \tag{29}$$

from which we have

$$R^2r^2 + x_4^2 = R^2/k \tag{30}$$

since R^2/k is a constant (at one instant of time t) it follows that

$$R^2 r\,dr + x_4 dx_4 = 0$$

or

$$(dx_4)^2 = \frac{kr^2r^2dr^2}{1 - kr^2}\ . \tag{31}$$

Using this result we can now rewrite dL^2, Eq.(25), as

$$dL^2 = R^2[dr^2 + r^2d\theta^2 + r^2\sin^2\theta\, d\varphi^2] + \frac{kR62r^2dr^2}{1 - kr^2}$$

$$= dx^2 + dy^2 + dz^2 + dx_4^2\ . \tag{32}$$

The right-hand side of the above equation is the expression for the line element of a 4-dimensional space with Cartesian coordinates x, y, z and x_4, and Eq.(29) implies that the variables are confined to be on the surface of a hypersphere with radius $R/\sqrt{k}$. Thus, for $k > 0$, the universe is a 3-space embedded in a (fictitious) 4-dimensional Euclidean space. The circumference of a circle in the "spherical" coordinates r, θ, φ is equal to $2\pi r$, and the surface of a sphere to $4\pi r^2$. But the "radius" of a circle (or sphere) is equal to

$$\int_0^r \frac{dr}{\sqrt{1 - kr^2}} = \frac{1}{\sqrt{k}} \sin^{-1}(\sqrt{k}\,r) \tag{33}$$

i.e., is larger than r. Thus the ratio of circumference to radius in this space is less than 2π. The volume of this space is finite. To show this, let us introduce in place of the coordinate r the "angle" variable

$$\sin\chi = \sqrt{k}\,r \tag{34}$$

the range of χ goes between the limits 0 and χ, which is found from the restriction $0 \leq \sqrt{k}\,r \leq 1$. In terms of this new variable, dL^2 takes the new form

$$dL^2 = a^2\left[d\chi^2 + \sin^2\chi(d\theta^2 + \sin^2\theta\, d\varphi^2)\right] \tag{35}$$

where $a^2 = R^2/k$. The coordinate χ determines the distance from the origin, given by $a\chi$, and the total volume of space is equal to

$$V = \int_0^{2\pi}\int_0^{\pi}\int_0^{\pi} a^3 \sin^2\chi \sin\theta\, d\chi\, d\theta\, d\varphi = 2\pi^2 a^3 = 2\pi^2 k^{-3/2} r^3 \,. \tag{36}$$

Thus a space of positive curvature turns out to be closed on itself. Its volume is finite though it has no boundaries.

For a space with $k < 0$ (i.e., having a constant negative curvature), the element of length has, in coordinates r, θ, and φ, the form

$$dL^2 = R^2(t)\left[\frac{dr^2}{1 + |k| r^2} + r^2(d\theta^2 + \sin^2\theta\, d\varphi^2)\right] \tag{37}$$

where the coordinate r can go through all values from 0 to ∞. The ratio of the circumference of a circle to its radius is now greater than 2π.

Corresponding to Eq.(34) for the angle variable χ, we now have

$$\sinh\chi = |k|^{1/2} r; \qquad 0 \leq \chi < \infty . \tag{38}$$

Then

$$dL^2 = (R^2/|k|)[d\chi^2 + \sinh^2\chi\,(d\theta^2 + \sin^2\theta\, d\varphi^2)] \tag{39}$$

and the surface of a sphere is now equal to

$$4\pi^2 r^2 = 4\pi^2 (R^2/|k|) \sinh\chi \tag{40}$$

and we see that as we move away from the origin (increasing χ), it increases without limit. The volume of this space is clearly infinite. Thus a universe of negative curvature is an open universe.

For a space with k = 0, the element of length reduces to

$$dL^2 = R^2(t)[dr^2 + r^2(d\theta^2 + \sin^2\theta\, d\varphi^2)] \tag{41}$$

the space is, clearly, Euclidean; the volume is again infinite.

Finally, we emphasize that the Robertson-Walker metric is a consequence only of the symmetry of 3-dimensional position space, expressed in a 4-dimensional language. The number of unknowns in the metric tensor is reduced from 10 to the single function, function R(t), and the discrete parameter k. Now let us see how simply the function R(t) describes the expansion of the universe. Specifically we will discuss some of the important observational features of a typical Robertson-Walker spacetime. These features show how a non-Euclidean geometry can differ substantially from conclusions based on naive Euclidean concepts.

The way in which the scale-factor R(t) varies with time has to be determined by substituting the Robertson-Walker metric into Einstein's field equations. We will defer this task to the next chapter, and we shall learn there that a variety of model universes can be introduced, characterized by different values of the scale-factor R(t).

9-5 Cosmic kinematics

(A) Hubble's law

Let us first try to understand how Hubble's law is accounted for by the Robertson-Walker model. We begin by recalling that each galaxy has similar coordinates (r, θ, φ). Now imagine for simplicity that our Galaxy, which is approximately co-moving, lies at the spatial origin $r = 0$, and consider another galaxy at r, say. Our choice of spatial origin does not place ourselves at the center of the universe. Because of the assumption of homogeneity, any galaxy could be chosen to have $r = 0$. Our particular choice is simply for convenience.

The proper distance of the distant galaxy from us at a given cosmic time t is D_p, given by the Robertson-Walker metric as

$$D_p = R(t)\int_0^r \frac{dr}{\sqrt{1 - kr^2}} . \tag{42}$$

The above integral is elementary and the result is

$$D_p = \begin{cases} R(t)\sin^{-1}r & (k = 1) \\ R(t)r & (k = 0) \\ R(t)\sinh^{-1}r & (k = -1) . \end{cases} \tag{42a}$$

Thus the proper distance is proportional to the scale-factor R(t) which changes with time. The proper velocity v_p is obtained by differentiating D_p with respect to t, and remembering that r remains constant because it is a co-moving coordinate:

$$v_p = \dot{D}_p = \dot{R}(t)\int_0^r \frac{dr}{\sqrt{1 - kr^2}} = \frac{\dot{R}(t)}{R(t)} D_p \tag{43}$$

where dots denote differentiation with respect to time. This says that at any given cosmic time t the speed of a galaxy relative to us is proportional to its proper distance from us. But this is simply Hubble's law, with the Hubble constant H given by

$$H(t) = \frac{\dot{R}(t)}{R(t)} . \tag{44}$$

We see that in general the Hubble constant H depends on time; what astronomers measure is the value H_0 of the Hubble constant at the present epoch $t = t_0$.

The proper distance D_p is not a measurable quantity, however, red shifts, apparent magnitude, number counts etc., are measurable. We shall discuss these below, and we shall see that they involve only the scale factor R(t).

(B) Red shifts

Consider a distant galaxy at $r = r_e$ emitting light towards us (at r = 0). Let us denote by t_0 the present epoch of observation (Figure 3). Two wave crests reaching us at t_0 and $t_0 + \Delta t_0$ were emitted at t_e and $t_e + \Delta t_e$ from the distant galaxy. Since light waves travel radially inward along a null geodesic, we assume that only r and t change along the null geodesic. Next we recall that a first integral of the null geodesic equation is simply ds = 0. For the Robertson-Walker line element this gives us

$$0 = c^2 \Delta t^2 - R^2(t) \Delta r^2 / (1 - kr^2)$$

or

$$c\,dt = \pm \frac{R(t)\,dr}{\sqrt{1 - kr^2}}$$

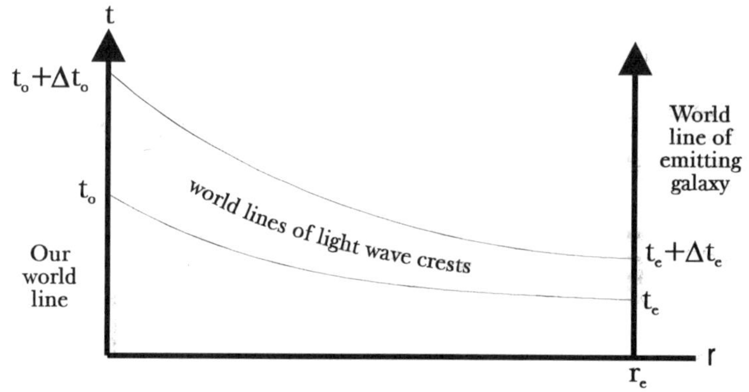

Figure 3: World lines for calculating cosmic red shifts.

Since r decreases as t increases along this null geodesic, we take the "-" sign in the above equation. Following this we have

$$\int_{t_e}^{t_0} \frac{dt}{R(t)} = \frac{1}{c}\int_0^{r_e} \frac{dr}{\sqrt{1 - kr^2}}$$

$$\int_{t_e+\Delta t_e}^{t_0+\Delta t_0} \frac{dt}{R(t)} = \frac{1}{c}\int_0^{r_e} \frac{dr}{\sqrt{1 - kr^2}} .$$

For all kinds of electromagnetic radiation received from galaxies, Δt_e and Δt_0 are tiny fractions of a second, and over such small time intervals R(t) remains effectively unchanged; therefore

$$\Delta t_0/R(t_0) - \Delta t_e/R(t_e) = 0 \quad \Rightarrow \quad \Delta t_0/\Delta t_e = R(t_0)/R(t_e) .$$

Now, the observed and emitted wavelengths λ_0 and λ_e are related to the periods Δt_0 and Δt_e by

$$\lambda_0 = c\Delta t_0 ; \qquad \lambda_e = c\Delta t_e$$

so that the red shift z, defined by the relation $z = (\lambda_0 - \lambda_e)/\lambda_e$, can be written in terms of $R(t_0)$ as

$$z = R(t_0)/R(t_e) - 1 . \tag{45}$$

In an expanding universe, $R(t_0) > R(t_e)$, so that z is positive, as observed.

It is worth emphasizing that our derivation above shows that the redshift effect arises from the passage of light through a non-Euclidean spacetime. It does not arise from the Doppler effect. We will refer to the present redshift as cosmological redshift.

If the scale-factor at the present epoch $R(t_0)$ is set equal to 1, then

$$1 + z = 1/R(t_e) . \tag{45a}$$

Redshift is simply a measure of the scale factor of the universe at t_e

(i.e., when the source emitted its radiation). For example, when we observe a galaxy with $z = 1$, the scale-factor of the universe at t_e was $R(t_e) = 0.5$, i.e. the universe was half its present size. But we have no information about when the light was emitted. If we did, we could determine the function $R(t)$ directly from observation. We have to have some theory of cosmic dynamics in order to determine $R(t)$. This is the subject of the next chapter.

We end this section with a discussion on a useful mathematical expansion. Most observed cosmological red shifts are rather small, so that t_e is (cosmologically speaking) not much earlier than t_0. We therefore expand $R(t_e)$ about t_0:

$$\begin{aligned} R(t_e) &= R(t_0) + (t_e - t_0)\dot{R}(t_0) + \tfrac{1}{2}(t_e - t_0)^2 \ddot{R}(t_0) + \ldots \\ &= R(t_0)\left[1 + H_0(t_e - t_0) - \tfrac{1}{2} q_0 H_0^2 (t_e - t_0)^2 + \ldots\right] \end{aligned} \qquad (46)$$

where H_0 is the present value of Hubble constant, i.e.

$$H_0 = \dot{R}(t_0)/R(t_0) \qquad (47)$$

and q_0 is the dimensionless <u>deceleration parameter</u>

$$q_0 = -\ddot{R}(t_0)R(t_0)/\dot{R}^2(t_0) = -\ddot{R}(t_0)/(R(t_0)H_0^2) \qquad (48)$$

Clearly q_0 is positive if the expansion of the universe is slowing down, i.e. if $\ddot{R}$ is negative. The redshift formula (45) can now be expanded in powers of $t_0 - t_e$:

$$z = H_0(t_0 - t_e) + (1 + q_0)H_0^2(t_0 - t_e)^2 + \ldots\ldots \qquad (49)$$

and, conversely, the time of light travel $t_0 - t_e$ may be expanded as a function of z:

$$t_0 - t_e = \frac{1}{H_0}\left[z - \left(1 + \frac{1}{2}q_0\right)z^2 + \ldots\right] \qquad (50)$$

These formulae are very useful, but it should not be forgotten that they are only valid for small z.

(C) Angular size

We now calculate how the apparent size of an object varies with distance from us. Consider an object of linear size (diameter) d with radial coordinate $r = r_1$. What angle does this length d subtend at our location $(r = 0)$? To answer this question, we consider two neighboring null

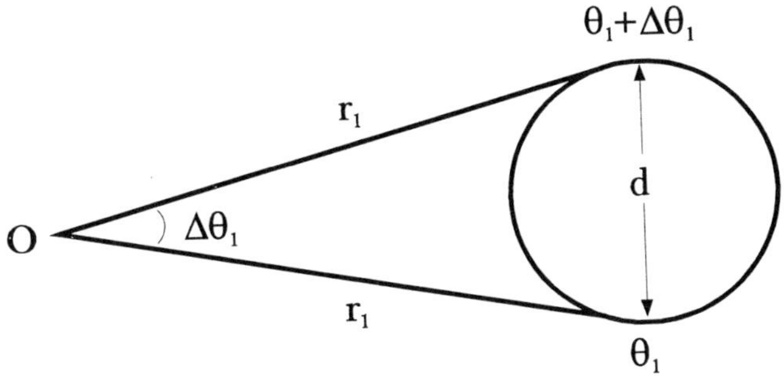

Figure 4: Angular size.

geodesics (light rays) from the two points A and B at the extremities of the object directed toward us. Point A has the coordinates $(r_1, \theta_1, \varphi_1)$ and B has the coordinates $(r_1, \theta_1 + \Delta\theta_1, \varphi_1)$. From the Robertson-Walker line element the proper distance between A and B is

$$ds^2 = -R^2(t_1)r_1^2(\Delta\theta_1)^2 = -d^2$$

since in the rest frame of the object the space-like separation AB = d. Thus

$$\Delta\theta_1 = \frac{d}{r_1 R(t_1)} = \frac{d(1+z)}{r_1 R(t_1)} \qquad (51)$$

which is the angle the length d subtended at our location.

It is interesting to note that as we are looking at more and more remote galaxies, we are also looking back into time. This means that r_1 increases, while t_1 decreases. Now $R(t_1)$ decreases as t_1 decreases, so it is not obvious that $r_1R(t_1)$ should get progressively larger as we look at more remote galaxies. We need to know how fast $R(t_1)$ decreases as r_1 increases. All that we can say at present is that the observed angular size of galaxies need not be a monotonic decreasing function of their distance, as it would be in the Euclidean universe.

(D) Apparent luminosity

The apparent luminosity of a galaxy or a star is a very important quantity for astronomers. Let us now see how the apparent luminosity is related to the intrinsic luminosity of the galaxy and its distance from us in the expanding universe described by the Robertson-Walker spacetime.

Suppose a distant galaxy G_1 emitted total energy L in unit time at the epoch t_1 and we receive it at the present epoch t_0. The red shift z of the galaxy is given by Eq.(45), $1 + z = R(t_0)/R(t_e)$. If the intensity distribution of G_1 over wavelengths λ is given by a normalized function $I(\lambda)$, then

$$dL = LI(\lambda)d\lambda \tag{52}$$

is the energy emitted by G_1 per unit time over the bandwidth $(\lambda, \lambda + d\lambda)$. We can use frequencies instead, in which case the intensity function $J(\nu)$ is related to $I(\lambda)$ by

$$J(\nu)|d\nu| = I(\lambda)|d\lambda|$$

$$cJ(\nu) = \lambda^2 I(\lambda); \qquad (c = \lambda\nu) . \tag{53}$$

If the light is emitted isotropically by G_1, by the time light reaches us it is distributed uniformly across a sphere of coordinate r_1 centered at G_1 (Figure 5).

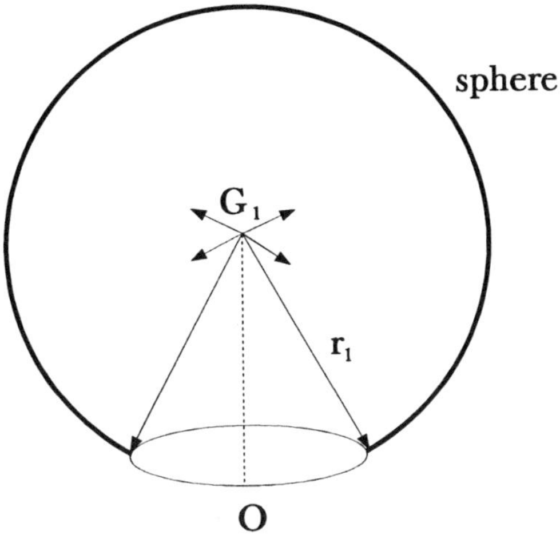

Figure 5: The radiation emitted by galaxy G_1.

The total surface area of this sphere is $4\pi r_1{}^2 R^2(t_0)$. It is easy to check this. Set t = constant, r = constant, and the Robertson-Walker metric reduces to the following simple form

$$ds^2 = -r^2 R^2(d\theta^2 + \sin^2\theta \, d\varphi^2) \ .$$

This is the line element on the surface of a Euclidean sphere of radius rR. Hence the total surface area at time t_0 is

$$4\pi r_1^2 R(t_0) \ .$$

Let $F(\lambda_0)\Delta\lambda_0$ denote the light received per unit time by us across unit proper area held perpendicular to the line of sight to G_1 and over a bandwidth $(\lambda_0, \lambda_0 + \Delta\lambda_0)$. Because of redshift, the arriving light with wavelengths in the range

$$\lambda_0, \ \lambda_0 + \Delta\lambda_0$$

left G_1 in the wavelength range

$$\left[\lambda_0/(1 + z), \ (\lambda_0 + \Delta\lambda_0)/(1 + z)\right]$$

and the total amount of energy that leaves G_1 between the epoch t_1 and $t_1 + \Delta t_1$ is

$$LI\left(\frac{\lambda_0}{1+z}\right)\frac{\Delta\lambda_0}{1+z}\,\Delta t_1$$

How many photons carry the above quantity of energy? For a small enough bandwidth, we may assume that a typical photon had, at emission, the wavelength $\lambda_0/(1+z)$ or a frequency $(1+z)c/\lambda_0$, hence an energy equal to $(1+z)ch/\lambda_0$, where h is the Planck constant. Therefore the required number of photons is

$$\delta N = LI\left(\frac{\lambda_0}{1+z}\right)\frac{\Delta\lambda_0}{1+z}\,\frac{\Delta t_1}{(1+z)ch/\lambda_0}$$

$$= \frac{L\lambda_0}{ch}\,\frac{1}{(1+z)^2}\,I\left(\frac{\lambda_0}{1+z}\right)\Delta\lambda_0\Delta t_1$$

At the epoch of reception, these photons are distributed across a surface area of $4\pi r_1{}^2R^2(t_0)$ and are received over a time interval $(t_0,\ t_0+\Delta t_0)$. Thus the number of photons received by us per unit time and per area held normal to the line of sight is given by

$$\frac{\delta N}{4\pi r_1^2 R^2(t_0)\,\Delta t_0}$$

and so each photon now has energy ch/λ_0. Multiplying by this factor, we obtain $F(\lambda_0)\Delta\lambda_0$:

$$F(\lambda_0)\Delta\lambda_0 = L\frac{1}{(1+z)^2}\frac{\Delta t_1}{\Delta t_0}I\left(\frac{\lambda_0}{1+z}\right)\frac{1}{4\pi r_1^2R^2(t_0)}\Delta\lambda_0\ .$$

But

$$\frac{\Delta t_1}{\Delta t_o} = \frac{R(t_1)}{R(t_o)} = (1+z)^{-1}\ .$$

Thus we finally obtain the flux density

$$F(\lambda_0) = \frac{LI(\lambda_0/(1 + z))}{(1 + z)^3 4\pi r_1^2 R^2(t_0)} . \tag{54}$$

In terms of frequencies we have

$$F(\nu_0) = \frac{LJ(\nu_0(1 + z))}{(1 + z) 4\pi r_1^2 R^2(t_0)} . \tag{55}$$

The optical astronomers use $F(\lambda_0)$, and the radio astronomers prefer to use $F(\nu_0)$. Integration of $F(\lambda_0)$ over all wavelengths gives the apparent (bolometric) luminosity of the galaxy. It is a common practice in optical astronomy to express apparent and intrinsic luminosities in terms of apparent and absolute magnitudes m and M. It is outside our syllabus to pursue this distinction.

(E) Galaxy counts

Galaxy counts may give indications that spacetime geometry is non-Euclidean. Let us suppose that at any epoch t there are n(t) galaxies in a unit proper coordinate volume. Although the galaxies individually retain the same coordinates (r, θ, φ), the proper separation between them at any epoch increases with epoch according to the scale-factor R(t). Accordingly, the proper volume of any region bounded by such galaxies increases as $R^3(t)$.

The number of galaxies seen in the radial coordinate range r to r + dr is

$$dN = \frac{4\pi r^2 dr}{\sqrt{1 - kr^2}} n(t) . \tag{56}$$

Note that when we observe galaxies at radial coordinates between r and r + dr, we see them in time range t to t + dt. In Robertson-Walker spacetime, t is related to r through the following relation

$$\int_t^{t_0} \frac{c\,dt'}{R(t')} = \int_0^r \frac{dr'}{\sqrt{1 - kr'^2}} .$$

Integration of the right-hand side of Eq.(56) from 0 to r_1 gives the number of galaxies out to r_1:

$$N(r_1) = \int_0^{r_1} \frac{4\pi r^2 n(t)dr}{\sqrt{1 - kr^2}} .$$

If we take n(t) = constant (this is not true for steady-state theory, as new galaxies can be formed), the integral can then be explicitly evaluated. $N(r_1)$ increases faster than the Euclidean form ($\propto r_1^3$) in a closed universe (k = 1) and slower in open universes (k = -1).

Problems

(9-1) Show that the Newtonian gravity does not admit a cosmology which is isotropic, homogeneous, and static.

(9-2) Show that an empty (containing no matter), isotropic spacetime is a flat Minkowski space.

(9-3) A celestial object emits black body radiation of temperature T in its own rest frame; we see it at a redshift Z and subtending a solid angle Ω. What flux do we measure?
(To solve this problem, we need Liouville's theorem which is not covered in the text. This theorem states that the number density of photons in phase space is a constant along a light ray.)

Chapter 10

Cosmic Dynamics
(Non-static Models of the Universe)

10-1 Dynamical equations of cosmology

The uniform model universes based on the Robertson-Walker line element are characterized by their scale factors R(t) and the curvature index k. But we have so far been concerned with the symmetry properties of the universe and the results that follow therefrom. The work covered in Chapter 9 did not tell us how the scale factor R(t) varies with time t, thus we do not know the rate at which the universe expands as given by the scale factor R(t). We also do not know whether the universe is open or closed as indicated by the curvature parameter k. To find answers to these questions we need a dynamic theory, i.e., to combine the isotropic, homogeneous Robertson-Walker line element with Einstein's field equations. This procedure will give us the dynamical equations satisfied by the scale factor R(t).

We first compute the Einstein tensor. If we label our coordinates according as $x^0 = ct$, $x^1 = r$, $x^2 = \theta$, $x^3 = \varphi$, then the non-zero components of $g_{\mu\nu}$ and $g^{\mu\nu}$ are:

$$g_{00} = 1 = g^{00}, \quad g_{11} = -\frac{R^2}{1 - kr^2} = (g^{11})^{-1}, \quad g_{22} = -R^2r^2 = (g^{22})^{-1}$$

$$g_{33} = -R^2r^2\sin^2\theta = (g^{33})^{-1}, \qquad \sqrt{-g} = \frac{R^3r^2\sin\theta}{\sqrt{1 - kr^2}} \,. \tag{1}$$

The non-zero components of $\Gamma^{\mu}_{\nu\alpha}$ are then as follows:

$$\Gamma^1_{01} = \Gamma^2_{02} = \Gamma^3_{03} = \frac{1}{c}\frac{\dot{R}}{R}, \qquad \left(\Gamma^i_{0j} = -\frac{1}{2}\frac{\partial}{\partial t}g_{ij}\right)$$

$$\Gamma^0_{11} = \frac{R\dot{R}}{c(1 - kr^2)}, \quad \Gamma^0_{22} = \frac{r^2R\dot{R}}{c}, \quad \Gamma^0_{33} = \frac{S\dot{R}r^2\sin^2\theta}{c}; \quad \left(\Gamma^0_{ij} = -\frac{1}{2}\frac{\partial}{\partial t}g_{ij}\right)$$

$$\Gamma^1_{11} = \frac{kr}{1 - kr^2}, \quad \Gamma^1_{22} = -r(1 - kr^2), \quad \Gamma^1_{33} = -r(1 - kr^2)\sin^2\theta \tag{2}$$

$$\Gamma^2_{12} = \Gamma^3_{13} = \frac{1}{r}, \quad \Gamma^2_{33} = -\sin\theta\cos\theta, \quad \Gamma^3_{23} = \cot\theta\ .$$

A dot, as before, denotes differentiation with respect to time. We now calculate the Ricci tensor which may be put in the following form:

$$R_{\mu\nu} = \frac{\partial^2 \ln\sqrt{-g}}{\partial x^\mu \partial x^\nu} - \frac{\partial\Gamma^\lambda_{\mu\nu}}{\partial x^\lambda} + \Gamma^\beta_{\mu\alpha}\Gamma^\alpha_{\nu\beta} - \Gamma^\lambda_{\mu\nu}\frac{\partial \ln\sqrt{-g}}{\partial x^\lambda}\ .$$

Straightforward but tedious calculation gives the following non-zero components of the Ricci tensor:

$$R^0_0 = \frac{3}{c^2}\frac{\ddot{R}}{R}, \quad R^1_1 = R^2_2 = R^3_3 = \frac{1}{c^2}\left(\frac{\ddot{R}}{R} + \frac{2\dot{R}^2 + 2kc^2}{R^2}\right). \tag{3}$$

From these we get the scalar curvature S

$$S = R^k_k = \frac{6}{c^2}\left(\frac{\ddot{R}}{R} + \frac{\dot{R}^2 + kc^2}{R^2}\right) \tag{4}$$

and hence the Einstein tensor G^μ_μ:

$$G^1_1 = R^1_1 - \frac{1}{2}S = -\frac{1}{c^2}\left(2\frac{\ddot{R}}{R} + \frac{\dot{R}^2 + kc^2}{R^2}\right) = G^2_2 = G^3_3 \tag{5}$$

$$G^0_0 = R^0_0 - \frac{1}{2}S = -\frac{3}{c^2}\left(\frac{\dot{R}^2 + kc^2}{R^2}\right). \tag{6}$$

We now apply the Einstein equations:

$$R^{\nu}_{\mu} - \frac{1}{2}\delta^{\nu}_{\mu}R^{\nu}_{\mu} = \frac{8\pi G}{c^4}T^{\nu}_{\mu}.$$

The "time-time" component (i.e., $\mu = 0$, $\nu = 0$) gives us

$$2\frac{\ddot{R}}{R} + \frac{\dot{R}^2 + kc^2}{R^2} = \frac{8\pi G}{c^2}T^1_1 = \frac{8\pi G}{c^2}T^2_2 = \frac{8\pi G}{c^2}T^3_3 \tag{7}$$

while the "space-space" component gives us

$$\frac{\dot{R}^2 + kc^2}{R^2} = \frac{8\pi G}{3c^2}T^0_0. \tag{8}$$

Note that the three non-trivial spatial equations are equivalent. This is essentially due to the homogeneity and isotropy of the Robertson-Walker line element.

We now need the energy tensor T^{ν}_{μ} to describe the cosmic fluid. As we idealize the universe, and model it by a simple macroscopic ideal fluid, devoid of viscous and heat-conductive properties, its energy tensor is then that of a perfect fluid, so

$$T^{\nu}_{\mu} = -p\delta^{\nu}_{\mu} + (p + \rho)U^{\nu}U_{\mu}; \qquad U_{\mu}U^{\mu} = 1 \tag{9}$$

where p is the pressure and ρ is the energy-density of the cosmic fluid, and U_{μ} is the (covariant) world velocity of the fluid particles (galaxies). In the co-moving frame that we have chosen, the fluid is at rest, accordingly

$$U^i = (1, 0, 0, 0) \tag{10}$$

and

$$T_0^0 = \rho c^2, \quad T_1^1 = T_2^2 = T_3^3 = -p \; . \tag{11}$$

Substituting these into Eqs.(7) and (8) we obtain

$$2\frac{\ddot{R}}{R} + \frac{\dot{R}^2 + kc^2}{R^2} = -\frac{8\pi G}{c^2}p \tag{12}$$

and

$$\frac{\dot{R}^2 + kc^2}{R^2} = \frac{8\pi G}{3}\rho \; . \tag{13}$$

Finally, we have the equations of motion of the fluid particles

$$T^{\nu}_{\lambda;\nu} = 0 \; . \tag{14}$$

Writing this out in full, we have

$$\frac{\partial p}{\partial x^{\lambda}} - \frac{\partial}{\partial x^{\nu}}\left[(p + \rho)U^{\nu}U_{\lambda}\right] - \Gamma^{\nu}_{\sigma\nu}(p + \rho)U^{\sigma}U_{\lambda} + \Gamma^{\sigma}_{\nu\lambda}(p + \rho)U^{\nu}U_{\sigma} = 0 \; .$$

With the aid of Eqs.(2) and (3) the above equation reduces to

$$\frac{\partial p}{\partial x^{\lambda}} - \frac{d}{dt}\left[(p + \rho)U_{\lambda}\right] - 3(p + \rho)U_{\lambda}\frac{\dot{R}}{R} = 0 \; .$$

The components $\lambda = 1, 2, 3$ of the above equation give the trivial equation: $0 = 0$; and the component $\lambda = 0$ gives the non-trivial equation:

$$\dot{\rho} + 3(p + \rho)\frac{\dot{R}}{R} = 0 \; . \tag{15}$$

The three equations (12), (13) and (16) are not all independent; for instance, taking a derivative of Eq.(13) and using Eq.(16) we easily produce Eq.(12). Thus, it is sufficient to retain any two of these three equations. We shall keep Eqs.(12) and (13) and refer to them as the Friedmann equations, after A. Friedmann.

Now in our smooth fluid approximation a velocity field like Eq.(10) represents an orderly motion with no pressure. That is, we have in this case the system of galaxies behaving like (incoherent) dust. With this approximation, Eq.(16) becomes

$$\frac{d}{dt}(\rho R^3) = 0$$

which integrates to

$$\rho R^3 = \text{constant} = \rho_0 R_0^3 \tag{16}$$

ρ_0 and R_0 being the values of ρ and R at the present epoch. This is the so-called Friedmann integral, which shows that matter density varies in time as R^{-3}, and the quantity of matter contained in a co-moving volume-element is constant during the expansion. This discussion is relevant to a description of a matter-dominated epoch. For a radiation-dominated epoch, pressure cannot be ignored and it is related to density by $p = c^2/3$.

In the present chapter we will consider the matter-dominated epochs, and so the two Friedmann equations become:

$$2\frac{\ddot{R}}{R} + \frac{\dot{R}^2 + kc^2}{R^2} = 0 \tag{17}$$

$$\frac{\dot{R}^2 + kc^2}{R^2} = \frac{8\pi G\rho_0}{3}\frac{R_0^3}{R^3}\ . \tag{18}$$

10-2 The solution of Friedmann's equations

We now consider the solutions of the Friedmann equations for the three cases k = 0, 1, -1. Before we consider the three cases separately, let us do some preparation work. First recall that the Hubble constant H(t) is defined by

$$H(t) \equiv \frac{\dot{R}(t)}{R(t)}$$

and we denote its present-day value by $H_0 = H(t_0)$. Now applying Eq.(18) to the present epoch and rewriting it in terms of H_0, we get

$$\frac{k}{R_0^2} = \frac{8\pi G \rho_0}{3c^2} - \frac{H_0^2}{c^2} = \frac{8\pi G}{3c^2}\left(\rho_0 - \frac{3H_0^2}{8\pi G}\right) . \tag{19}$$

Hence $k > 0$, $k = 0$ or $k < 0$ as $\rho_0 > \rho_c$, $\rho_0 = \rho_c$, or $\rho_0 < \rho_c$ respectively, where ρ_c is called the critical (or closure) density given by

$$\rho_c = 3H_0^2 / 8\pi G . \tag{20}$$

With the range of values of H_0 known today, we have

$$\rho_c = 2 \times 10^{-29} h_0^2 \text{ g cm}^{-3}$$

where h_0 lies in the range $0.5 < h_0 < 1$. We next introduce the deceleration parameter q(t) by the relation

$$\frac{\ddot{R}}{R} = -q(t)[H(t)]^2, \quad \text{or} \quad q(t) = -\frac{R\ddot{R}}{\dot{R}^2} \tag{21}$$

and q_0 denotes the present-day value of q(t). Using Eqs.(17) and (18) gives

$$q_0 = \frac{4\pi G\rho_0}{3H_0^2} = \frac{\rho_0}{2\rho_c} . \tag{22}$$

(A) Flat model (k = 0)

For $k = 0$, Eq.(19) gives $\rho_0 = \rho_c$ then Eq.(22) gives $q_0 = 1/2$. Returning to the Friedmann equation (18), which becomes

$$\dot{R}^2 = \frac{8\pi G\rho_0}{3}\frac{R_0^3}{R} = \frac{A^2}{R}; \quad A^2 = \frac{8\pi G\rho_0 R_0^3}{3} \tag{23}$$

integration gives

$$R(t) = (3A/2)^{2/3} t^{2/3} . \tag{24}$$

(The integration constant has been set equal to zero by assuming that R = 0 at t = 0.) We also get

$$t_0 = \frac{2}{3H_0} . \tag{25}$$

Figure 1 illustrates this solution. The present epoch t_0 is denoted by the point A on the t-axis. The ordinates at A gives the present value of the scale factor, PA = R_0. The present value of the Hubble constant H_0 is given by the ratio 1/AB, where B is the intersection point of the tangent to the R(t) curve at P with the t-axis. The age of the universe, represented by 0A, is 2/3 of the intercept AB. Note that $\dot{R} \rightarrow 0$ as $t \rightarrow \infty$.

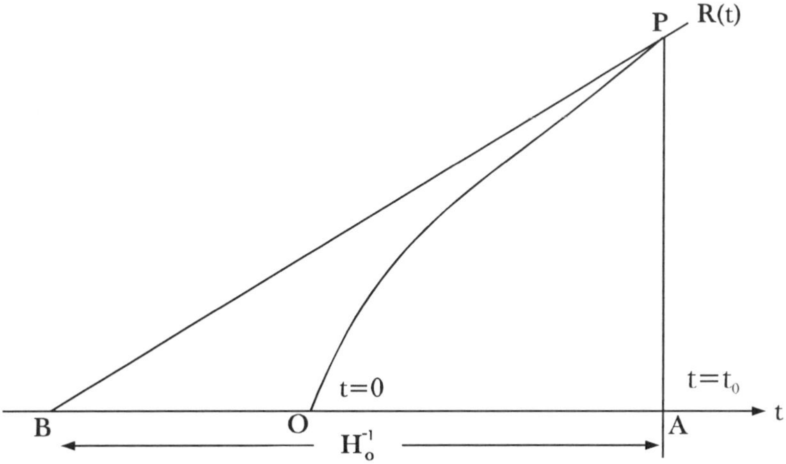

Figure 1: A schematic graph of R(t) as a function of t for the flat model (the Einstein-de Sitter model).

This model is also known as the Einstein-de Sitter model — it was given by Einstein and de Sutter in a joint paper in 1932.

(B) Closed model (k = 1)

For k = 1, Eq.(19) gives $\rho_0 > \rho_c$, then Eq.(22) gives $q_0 > 1/2$. Eqs.(17) and (18) become in this case

$$2\frac{\ddot{R}}{R} + \frac{\dot{R}^2 + c^2}{R^2} = 0 \tag{26}$$

$$\frac{\dot{R}^2 + c^2}{R^2} = \frac{A^2}{R^3} \tag{27}$$

where $A^2 = 8G_0R_0{}^3/3$. Now Eq.(27) gives

$$\frac{dR}{dt} = c\sqrt{\frac{B^2 - R}{R}}; \quad B^2 = \frac{A^2}{c^2}$$

from which we have

$$ct = \int_0^R \sqrt{\frac{R}{B^2 - R}}\, dR .$$

Make the substitution

$$R = B^2 \sin^2(\phi/2) = \frac{1}{2}B^2(1 - \cos\phi) .$$

The integral then becomes

$$ct = \frac{1}{2}B^2 \int (1 - \cos\phi)d\phi = \frac{1}{2}B^2(\phi - \sin\phi)$$

where we have taken $R = 0$ at $t = 0$ ($\phi = 0$). Now we have

$$R = \frac{1}{2}B^2(1 - \cos\phi), \qquad ct = \frac{1}{2}B^2(\phi - \sin\phi) \tag{28}$$

and these two equations give $R(t)$ via the parameter ϕ. The graph of $R(t)$ is a cycloid. In closed models, therefore, expansion is followed by contraction and R decreases to zero. The value $R = 0$ is reached when $\phi = 2\pi$; that is, when $t = t_L = B^2\pi/c$. Now B^2 can be expressed in terms of q_0 and H_0 as (to be shown later)

$$B^2 = \frac{2q_0}{(2q_0 - 1)^{3/2}} \frac{c}{H_0} \tag{29}$$

so

$$t = t_L = \frac{2\pi q_0}{(2q_0 - 1)^{3/2}} \frac{1}{H_0} . \tag{30}$$

The quantity t_L may be termed the "lifespan" of this universe. For $q_0 = 1$, $t_L = 2\pi/H_0$.

To derive Eq.(29), we apply Eqs.(17) and (18) to the present epoch and express them in terms of q_0 and H_0 as

$$\frac{c}{R_0^2} = (2q_0 - 1)H_0^2 \qquad (31)$$

$$\rho_0 = \left(H_0^2 + \frac{c^2}{R_0^2}\right)\frac{3}{8\pi G} = \frac{3H_0^2}{4\pi G}q_0 \ .$$

Substituting R_0 and ρ_0 from these two equations into B^2

$$B^2 = \frac{A^2}{c^2} = \frac{8\pi G\rho_0 R_0^3}{3c^2}$$

we obtain, after some straightforward manipulations, the desired result, namely Eq.(29).

Applying Eq.(28) to the present epoch we get

$$R_0 = \frac{1}{2}B^2(1 - \cos\phi_0) \ . \qquad (32)$$

Now substituting B^2 from Eq.(29) and R_0 from Eq.(31) into the above equation we get

$$\frac{1}{2}\frac{2q_0}{(2q_0 - 1)^{3/2}}\frac{c}{H_0}(1 - \cos\phi_0) = \frac{c}{H_0}\frac{1}{(2q_0 - 1)^{1/2}} \ .$$

From this we obtain, after simplification,

$$\cos\phi_0 = \frac{1 - q_0}{q_0} \ ; \qquad \sin\phi_0 = \frac{\sqrt{2q_0 - 1}}{q_0} \ .$$

We therefore get from Eq.(28)

$$t_0 = \frac{B^2}{2c}(\phi_0 - \sin\phi_0)$$

$$= \frac{q_0}{(2q_0 - 1)^{3/2}}\left[\cos^{-1}\left(\frac{1 - q_0}{q_0}\right) - \frac{\sqrt{2q_0 - 1}}{q_0}\right]\frac{1}{H_0} \ . \qquad (33)$$

For example, for $q_0 = 1$, we get

$$t_0 = \left(\frac{\pi}{2} - 1\right)H_0^{-1} \ . \qquad (34)$$

From Eq.(28) we see that R reaches a maximum value at $\phi = \pi$:

$$R = R_{max} = B^2 = \frac{2q_0}{(2q_0 - 1)^{3/2}} \frac{c}{H_0} \ . \tag{35}$$

Thus for $q_0 = 1$, the universe expands to twice its present size. Figure 2 illustrates the function R(t) for the closed models for a number of parameter values q_0. All curves have been scaled to touch at P, the present point, and they all have the common tangent PB. The intercept $AB = H_0^{-1}$. Notice that as q_0 increases the curves for R(t) intercept the past section of the t-line at points 0_1, 0_2, ... lying closer and closer to A, that is, the age of the universe is reduced if q_0 is increased.

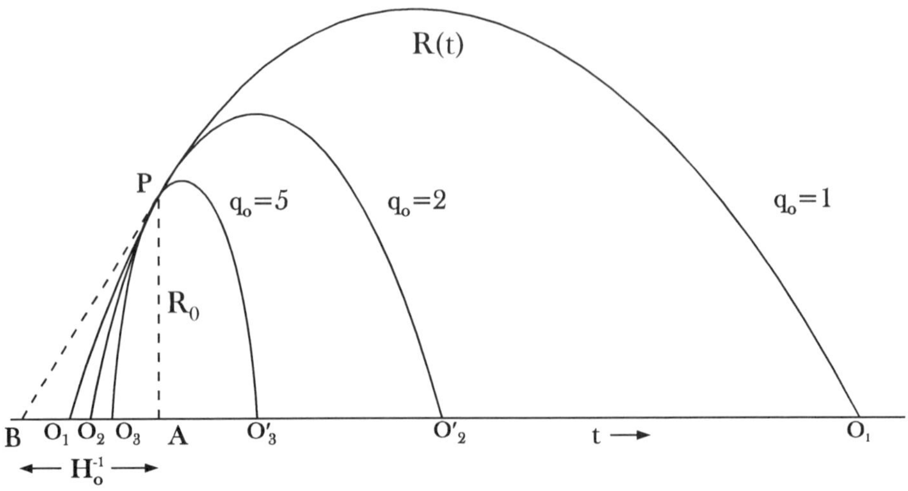

Figure 2: Closed model. The schematic R(t) curves for q_0 = 1, 2, 5.

(c) Open model (k = -1)

For k = -1, Eqs.(19) and (22) give $\rho_0 > \rho_c$, $q_0 < 1/2$; and Eqs.(17) and (18) become in this case

$$2\frac{\ddot{R}}{R} + \frac{\dot{R}^2 - c^2}{R^2} = 0 \tag{36}$$

$$\frac{\dot{R}^2 - c^2}{R^2} - \frac{8\pi G\rho_0 R_0^3}{3R^3} = 0 \tag{37}$$

Applying them to the present epoch, we can rewrite Eqs.(36) & (37) in terms of q_0 and H_0 as

$$\frac{c^2}{R_0^2} = (1 - 2q_0)H_0^2; \qquad \rho_0 = \frac{3H_0^2}{4\pi G}q_0 \,. \tag{38}$$

Eq.(37) gives

$$\dot{R}^2 = \frac{8\pi G\rho_0 R_0^3}{3}\frac{1}{R} + c^2 = c^2\left(1 + \frac{\alpha}{R}\right) \tag{39}$$

$$\alpha = \frac{8\pi G\rho_0 R_0^3}{3c^2} = \frac{2q_0}{(1 - 2q_0)^{3/2}}\frac{c}{H_0} \,. \tag{40}$$

From Eq.(39) we get

$$\frac{dR}{dt} = c\left(1 + \frac{\alpha}{R}\right)^{1/2}$$

so

$$ct = \int_0^R \sqrt{\frac{R}{\alpha + R}}\, dR \,.$$

Make the substitution

$$R = \alpha \sinh^2(\phi/2) = \frac{1}{2}\alpha(\cosh\phi - 1) \,. \tag{41}$$

Then the integral becomes

$$ct = \frac{1}{2}\int(\cosh\phi - 1)d\phi = \frac{1}{2}(\sinh\phi - \phi)$$

Again, as previous two cases, we have taken $R = 0$ at $t = 0$ ($\phi = 0$) Thus we have

$$R(t) = \frac{1}{2}\alpha(\cosh\phi - 1), \quad ct = \frac{1}{2}\alpha(\sinh\phi - \phi) \tag{42}$$

these give R(t) via the parameter ϕ. Since $R = \sinh\phi/(\cosh\phi - 1)$ so $R \rightarrow 1$ as ϕ (and hence t) $\rightarrow \infty$. Like the Einstein-de Sitter model, this model continues to expand forever.

It is easy to show, in a manner similar to the case k = 1, that the present value of ϕ is given by

$$\cosh\phi_0 = \frac{1 - q_0}{q_0}, \qquad \sinh\phi_0 = \frac{\sqrt{1 - 2q_0}}{q_0} \tag{43}$$

The present value of t is given by

$$t_0 = \frac{\alpha}{2c}(\sinh\phi_0 - \phi_0)$$

$$= \frac{q_0}{(1 - q_0)^{3/2}}\left[\frac{\sqrt{1 - 2q_0}}{q_0} - \ln\left(\frac{1 - q_0 + \sqrt{1 - 2q_0}}{Q_0}\right)\right]H_0^{-1} \tag{44}$$

The behavior of R(t) is illustrated in Figure 3. As in Figure 2, all curves have the same value of H_0 at P. The age of the universe is seen to increase as q_0 decreases, being maximum $\left(= H_0^{-1}\right)$ for $q_0 = 0$.

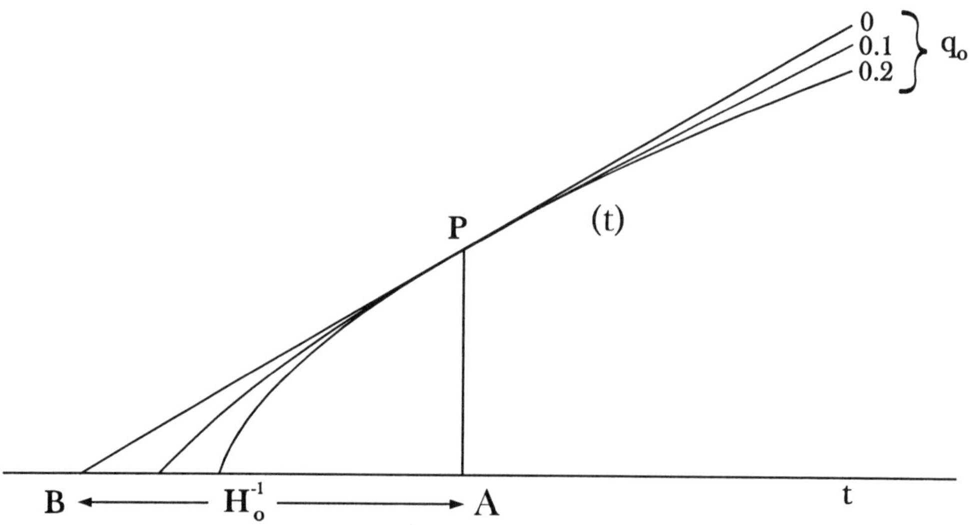

Figure 3: Open model. The schematic R(t) curves for q_0 = 0, 0.1, and 0.2.

10-3 Is the Universe open or closed?

The question naturally arises as to whether our universe is open or closed. Astronomical observations over the last fifty years or so have yielded estimates for H_0, q_0, and ρ_0. We shall not attempt a review of the methods employed, as they are extremely ingenious and complex. Recent estimates for H_0 put it at about $100h_0$ km s^{-1} Mpc^{-1} with h_0 in the range $0.5 < h_0 < 1$, and using Eq.(20) which gives a value for the critical density ρ_c of about $2 \times 10^{-26} h_0^2$ kg m^{-3}. Estimates for the deceleration parameter q_0 are more difficult to obtain, but the evidence suggests that $q_0 = 1$, in which case the universe is closed. But $q_0 = 1$ gives, via Eq.(22),

$$\rho_0 = 2\rho_c = 4 \times 10^{-26} h_0^2 \text{ kg m}^{-3}$$

which is much greater than the observed density of the universe. This discrepancy leads to the so-called "problem of missing mass".

The missing mass problem was investigated for the first time in 1930s by the Swiss astronomer Zwicky. He measured the mass of clusters of galaxies in two ways. Because there is a definite relation between the luminosity of a galaxy and its mass, Zwicky determined the mass of galaxies from their luminosity measurements. Then by adding up all the masses of the member galaxies, he obtained the total mass of the cluster. Mass of the cluster can also be determined by measuring the relative velocities among the galaxies, since the mean relative velocity is determined by the mass of the cluster. Zwicky discovered that the masses determined by these two methods differed greatly. For example, the dynamical mass of the Coma cluster is 400 times the luminosity mass! Zwicky believed that there must be a large amount of invisible matter within the cluster. Nothing was known at that time about what sort of matter contributing to this invisible mass, so it was called "missing mass".

Zwicky's bold conjecture was not well received. Since the 1950s, observational evidences supporting Zwicky's "missing mass" conjecture began to increase. The first decisive evidence appeared in 1978 and it came from the rotation curves of galaxies. The galactic rotation curves refer to the relation between the velocity of bodies revolving around (spiral) galaxies and their radius. For the solar system, a planet's velocity around the sun v and its radius r are related by Kepler's Third Law:

$$v \propto r^{-1/2} .$$

Thus, the further away a planet is from the sun, the smaller is its rotational velocity (Figure 4). Kepler's third law is applicable to motions around any massive central body. However, the observed rotational curve of a galaxy is completely different from Kepler's law: the rotating velocity of bodies is independent of their distance (Figure 5). This only possible interpretation of this result is that the space

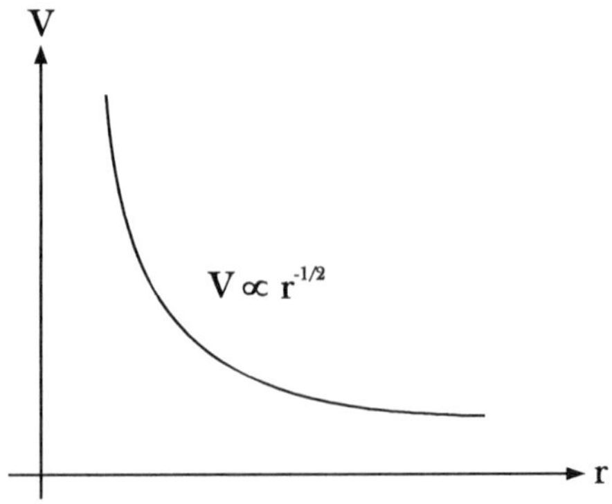

Figure 4: A planet's velocity around the sun (Kepler's Third Law).

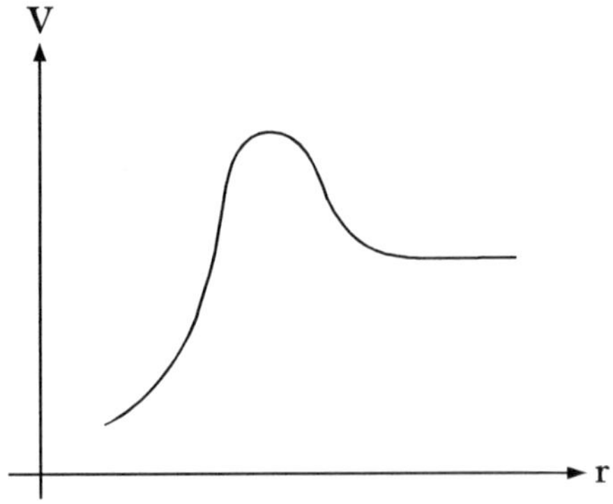

Figure 5: A schematic curve of the rotational velocity of a galaxy.

surrounding the galaxy is not empty; rather, it is a halo with a considerable mass. It does not emit light, so it is invisible. Today, it is generally accepted that up to 90 percent of the universe is made of invisible "dark matter".

What, then, is dark matter? Diffuse gas was the first candidate for consideration. Our galaxy contains many gas clouds. Could there be gaseous matter in large amounts in the intergalactic space? 21-cm observations show that the density of intergalactic hydrogen gas is less than 10^{-2} atoms per cubic centimeter — not enough to meet the deficit found by Zwicky. The result from optical methods is that the density of intergalactic hydrogen cannot exceed 10^{-12} atoms per cm^3. The optical methods can also detect other intergalactic atoms; the result has been entirely negative.

Yet, perhaps the intergalactic gas is in the ionized form. A high-temperature ionized gas will emit x-rays, and x-rays have indeed been found in clusters of galaxies. But the density found is far from enough to account for the deficit mass.

Recent experiments indicate that neutrinos may have a rest mass of $\sim 6 \times 10^{-32}$ g. If this is accurate, then neutrinos may just be the missing matter, because there are enormous numbers of neutrinos in the present universe. We can make a simple calculation for that number. We shall see in the next chapter that the annihilation of electrons and positrons in the early universe produced a sea of neutrinos and antineutrinos. At that time their numbers would have been comparable to the number of "cosmic" photons. In any volume R^3 that follows the expansion of the universe, the numbers of both photons and neutrinos would have been conserved (assuming negligible rates of net annihilations); so the number density of neutrinos and antineutrinos today must still be comparable to the number density of photons. Hence we can calculate the latter value and thereby get an estimate of the number density of neutrinos and antineutrinos.

The cosmic photons are from cosmic background radiation which is a relic radiation from the big bang. It will be shown in the next chapter that the cosmic background radiation has a black body spectrum at temperature $T = 2.7$ K. Now the energy density associated with black body radiation of temperature T is aT^4, and the average energy per photon is $\sim kT$. Thus, the number density of photons is equal to $aT^4/(kT) = aT^3/k$, or about 4.0×10^2 cm^{-3} for $T = 2.7$ K. Therefore if we take the current number density of neutrinos as 400 cm^{-3}, then its contribution to mass density of the universe is

$$6 \times 10^{-32} \times 400 = 2.4 \times 10^{-29} \text{ g cm}^{-3}$$

which is > ρ_c, thus it could make the universe finite and closed.

It is still an open question as to whether neutrinos have a rest mass or not. But discussion of neutrinos' rest mass has changed the perspective on the dark matter problem. Now it is generally regarded as an area to be opened up by the combined efforts of astrophysics and particle physics. Both theories of supergravity and supersymmetry predict the existence of many new particles, and these new particles have very weak interaction which cannot be detected in today's laboratories. Some physicists have speculated already that the dark matter in the universe may consist of just these particles.

Instead of trying to find all of the matter needed to close the universe, we can look for its gravitational effects. We can try to measure the actual slowing down of the expansion of the universe and see if $-\ddot{R}$ is large enough to stop the expansion. When we do this, we are determining the current value of the deceleration parameter q_0. From Eq.(21) when applied to the present epoch, we get

$$q_0 = -\ddot{R}(t_0)/H_0^2 \ .$$

As we do not measure $\ddot{R}(t_0)$, we would like to express $\ddot{R}(t_0)$ in terms of $H(t_0)$ and $\dot{H}(t_0)$. Now $H(t) = \dot{R}(t)/R(t)$, from which we have

$$\dot{R}(t) = H(t)R(t) \ .$$

Differentiating both sides with respect to time t gives

$$\ddot{R}(t) = H(t)\dot{R}(t) + \dot{H}(t)R(t) \ .$$

Setting $t = t_0$, and remembering that $R(t_0) = 1$, this becomes

$$\ddot{R}(t_0) = H_0^2 + \dot{H}(t_0) \ .$$

Substituting this back into the expression for q_0, we get

$$q_0 = -\left[\dot{H}(t_0)/H_0^2 + 1\right] \ . \qquad (45)$$

Eq.(45) says that if we can measure $H(t_0)$ and $\dot{H}(t_0)$ we can determine

q_0. How do we determine $\dot{H}(t_0)$? We can take advantage of the fact that when we look deeper and deeper into space, we are also looking farther and farther back into time. Thus, if we can determine H for objects that are, say, five billion light years away, we are really determining the value of H five billion years ago. If we include near and distant objects in a plot of Hubble's law, we will see deviations from a straight line. A sample result is shown in Figure 6. On the horizontal axis, we plot the brightness of a galaxy and on the vertical axis the red shift. The dashed line is an extrapolation of Hubble's law. If $q_0 = 1/2$, the universe barely manages to expand forever; if $q_0 > 1/2$, the universe is closed; if $q_0 < 1/2$, the universe is open.

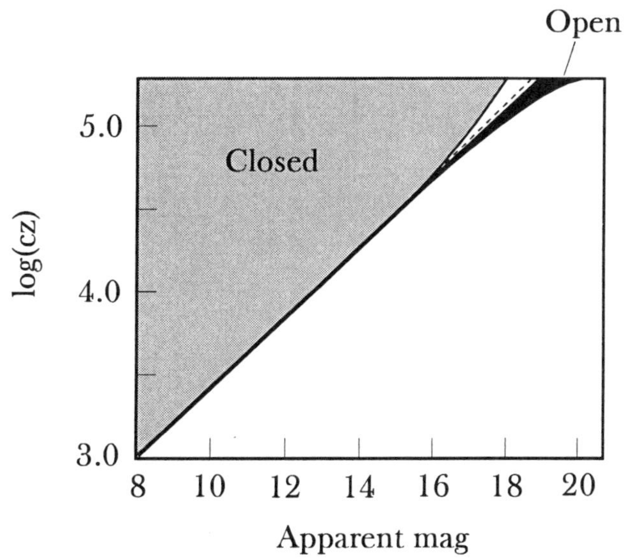

Figure 6: Hubble's law with deviations for distant objects.

In Figure 6, the dark region is open, and the shaded region is closed. But measuring $H(t_0)$ is not easy. The difficulty comes in the methods for measuring distances to distant objects.

Alternatively, we can determine k by surveying the universe on a large scale. One way of doing it is by radio source counts. If the universe is flat, the number of sources will go up as r^2. For a curved universe, as we look farther away, we should see deviation from the r^2 dependence.

10-4 Comment on Einstein's static model

When Einstein formulated the general theory of relativity in 1916, the universe was generally believed to be static. However, when Einstein applied his relativistic field equations to a "cosmic gas" of the kind we have been discussing, he found that if the density is not zero, the universe must necessarily be expanding or collapsing. To get around this problem, Einstein introduced into the field equations a cosmological term that provides a repulsion mechanism. The revised field equations read

$$R_{\mu\nu} - \frac{1}{2} g_{\mu\nu} R^{\alpha}_{\alpha} - \lambda g_{\mu\nu} = -\frac{8\pi G}{c^4} T_{\mu\nu}$$

where λ is constant, called the cosmological constant. The cosmological constant has to be extremely small, so as not to disturb the good agreement of the general-relativistic predictions with observations for the solar system. Since $R_{\mu\nu}$ contains second derivatives of $g_{\mu\nu}$, λ must have the dimensions of (distance)$^{-2}$. For λ to be small this distance must be very large. It is a cosmological distance, of the order of the radius of the universe.

After Hubble's detection of redshift in the 1920s and 1930s, Einstein withdrew the cosmological term, and later referred to it as the greatest blunder of his life. But it has not been so easy to eliminate the cosmological constant — it has survived to frustrate many theoretical physicists. According to the standard model of elementary particle physics, the cosmological constant is proportional to the vacuum energy density and it should be much greater than it is. If theories of elementary particles were correct, our universe would be dramatically different, with properties we would find both bizarre and unsettling. Could theories of elementary particles have somehow gone awry? No one can answer this question at present. An understanding of the disagreement could revolutionize fundamental physics.

10-5 Cosmology with Newtonian gravitation

We can learn a lot about the evolution of an expanding universe by applying Newtonian gravitation. Suppose we have a fluid with zero pressure moving according to Newton's laws of motion and gravitation. Such fluid is often called a Newtonian dust. We now consider a Newtonian dust of uniform density $\rho(t)$, which is in a state of uniform expansion, the only force on it being gravity. This means that the position vector $\vec{r}$ of a fluid particle at any time t is given by a relation of the form

$$\vec{r}(t) = R(t)\vec{c} \tag{46}$$

where $\vec{c}$ is a constant vector which is determined by the initial position of the fluid particle, and R(t) is the scale factor. Differentiation gives

$$\dot{\vec{r}}(t) = H(t)\vec{r}(t) \tag{47}$$

where

$$H(t) = \dot{R}(t)/R(t) \tag{48}$$

which is a Hubble constant associated with the scale factor R(t).

The Newtonian continuity equation

$$\frac{\partial \rho}{\partial t} + \nabla \cdot (\rho \dot{\vec{r}}) = 0$$

gives

$$\dot{\rho} + 3\rho H = 0$$

or

$$R\dot{\rho} + 3\dot{R}\rho = 0$$

which integrates to give

$$\rho R^3 = \text{constant} .$$

As in previous sections, let us indicate present-day values with a subscript zero, and write the above equation as

$$\rho R^3 = \rho_0 R_0^3 \tag{49}$$

which is the same as Eq.(16).

The assumption of isotropy implies that the universe appears to be spherically symmetric from any point. Thus any spherical volume evolves only under its own influence. The gravitational force exerted on the volume by material outside the volume sums to zero. If the volume in question has a radius r, and mass M(r), the equation of motion of a particle of mass m on the surface of the sphere as position **r** is

$$m\ddot{\vec{r}} = -\frac{GM(r)m\hat{r}}{r^2} \tag{50}$$

where $\hat{r}$ is a unit vector in the r-direction. Now the assumption of homogeneity means that the density ρ is the same everywhere, though it can change with time. The mass M(r) is then given by

$$M(r) = (4\pi/3)r^3\rho \ .$$

Substituting this and Eq.(46) into Eq.(50) gives

$$\ddot{R} = -(4\pi/3)G\rho R \ . \tag{51}$$

With the aid of Eq.(17), and since $R_0 = 1$, we have

$$\ddot{R} = -\frac{4\pi G\rho_0}{3}\frac{1}{R^2} \ . \tag{52}$$

Note that if ρ_0 is not zero, $\ddot{R}$ cannot be zero. Thus, a universe with matter cannot be static, it must be expanding or contracting.

Now let us multiply both sides of the equation of motion by $\dot{R}$ to give

$$\dot{R}\ddot{R} + \frac{(4\pi/3)GR\rho_0}{R^2}\dot{R} = 0 \ . \tag{53}$$

This becomes, using the fact that $d(R^2)/dt = 2\dot{R}R$,

$$\frac{1}{2}\frac{d}{dt}\dot{R}^2 + \frac{4\pi}{3}\frac{G\rho_0}{R^2}\frac{dR}{dt} = 0 \ . \tag{54}$$

Noting that $d(1/R)/dt = -(1/R^2)(dR/dt)$, we have

$$\frac{d}{dt}\left[\dot{R}^2 - \frac{8\pi G\rho_0}{3}\frac{1}{R}\right] = 0$$

which integrates to give

$$\dot{R}^2 - \frac{8\pi G\rho_0}{3}\frac{1}{R} = -k \qquad (55)$$

where k is a constant of integration. This is exactly the same as the relativistic equation (18), but there k is either ±1 or 0. We now rewrite Eq.(55) as

$$\frac{8\pi G\rho_0}{3} = (\dot{R}^2 + k)R \ .$$

Substituting into Eq.(52) gives

$$2\ddot{R} + \frac{\dot{R}^2 + k}{R} = 0 \ .$$

This is exactly the same as Eq.(17), except that k here is a constant of integration. In fact, if $k \neq 0$, there is no loss of generality in taking it to be ± 1, and we are therefore led to exactly the same three models for the evolution of the universe as we obtained in the preceding section.

Eq.(55) was obtained by integrating the equation of motion. Therefore we may regard it as the energy equation of the whole universe. We re-write it as

$$\dot{R}^2 - \frac{\alpha}{R} = -k$$

where we have set $\alpha = 8\pi G\rho_0/3$. We may regard $\dot{R}^2$ as a measure of its kinetic energy, $-\alpha/R$ as a measure of its gravitational potential energy, and -k as a measure of its total energy. If k = -1, the total energy is positive, and the universe has an excess of kinetic energy which allows it to keep expanding at an ultimately constant rate $\dot{R} \rightarrow 1$ as $t \rightarrow \infty$ (and so $R \rightarrow \infty$). If k = 0, the total energy is zero, and the kinetic energy is just sufficient to allow the universe to keep expanding, but at a decreasing rate ($\dot{R} \rightarrow 0$ as $t \rightarrow \infty$ and $R \rightarrow \infty$). If k = 1, the universe has insufficient kinetic energy for continued expansion. It expands until $R = A^2$ and $\dot{R} = 0$. After this, contraction will reverse the motion.

Many of the results of this section are similar to the general relativistic results we obtained in the previous section. This is because both depend on the fact that in a spherically symmetric distribution, matter outside a sphere has no effect on the evolution of the matter inside the sphere. But general relativity is essential for a proper treatment of cosmology, at least for the following reasons: (1) There are difficulties with Newtonian cosmology which our simple treatment obscures. (2) In relativity, pressure contributes to the total energy and hence to the gravitational field; Newtonian gravity is deficient in this respect. (3) Newtonian physics is inadequate in dealing with relativistic fluid particles (galaxies, etc.). (4) The problem of light propagation throughout the universe should be handled from a relativistic view-point.

Problems

(10-1) Check the expressions for the Ricci tensors and the Einstein tensors for the Robertson-Walker line element (i.e., Eqs.(3), (4), (5) and (6)).

(10-2) A galaxy is observed with redshift 0.69. How long did light take to travel from the galaxy to us if we assume that we are in the Einstein-de Sitter universe? ($H_0 = 100$ km/s·Mpc).

(10-3) From the equation $T^{\mu\nu}_{\ \ ;\mu} = 0$ obtain Eq.(15), and deduce that, if p = 0, then $\rho \propto 1/R^3$.

(10-4) Show that in the presence of the Einstein's λ term, Eqs.(7) and (8) are modified to the following:

$$2\frac{\ddot{R}}{R} + \frac{\dot{R}^2 + kc^2}{R^2} - \lambda c^2 = \frac{8\pi G}{c^2} T_1^{\ 1}$$

and

$$\frac{\dot{R}^2 + kc^2}{c^2} - \frac{1}{3}\lambda c^2 = \frac{8\pi G}{3c^2} T_0^{\ 0} .$$

Chapter 11

The Physics of the Early Universe

All the Friedmann models which were described in the previous chapter have the common feature that R = 0 at certain epochs (which we have chosen to label by t = 0). As we approach R = 0, the Hubble constant increases rapidly, becoming infinite at R = 0. This epoch therefore indicates violent activity and is given the name **big bang**. At this epoch the mathematical description of space-time geometry breaks down. R = 0 is also an insurmountable barrier to physicists: laws of physics break down also at R = 0. This doesn't mean that physics will never be able to explain the basic problem in cosmology. But we should move forward with caution and avoid over-confident in simple extrapolations based on our obviously limited physical knowledge at present. In this chapter we will concentrate on the physics of the early universe. The early universe here means the radiation-dominated era. To others, early universe may mean the epoch of the first three minutes after the birth of the universe.

11-1 Cosmic background radiation

We have no way of knowing directly the physical conditions of the big bang epoch. We can only look for relics of the big bang. Pioneering work in this field was done by George Gamow in the mid 1940s. Gamow was interested in the origin of elements. Starting from the basic building blocks of protons and neutrons, he attempted to describe the formation of nuclei through the fusion process. Astrophysicists already knew by the 1940s that such fusion processes operate inside stars, where the necessary conditions of high temperature and density were known to exist. Gamow pointed out that similar conditions must have existed in a typical Friedmann universe soon after the big band.

We know from Eq.(10-16) that the mass density $\rho(t)$ of the matter in the universe was very high at small values of R:

$$\rho(t) = \frac{\rho_0 R_0^3}{R^3(t)} . \tag{1}$$

What about temperature T? A simple calculation shows that the temperature was also very high at small R. The early universe contains radiation in the form of photons moving in all directions with high frequencies. Expansion causes the energy density $\epsilon_r(t)$ of this radiation to decrease as $R^{-4}(t)$ for two reasons: first, the number density of photons decreases as $R^{-3}(t)$ (because volumes expand as $R^3(t)$); and second, the energy of individual photons decreases as $R^{-1}(t)$ (because of the redshift in frequency). Therefore $\epsilon_r(t)$ decreases as $R^{-4}(t)$, and

$$\epsilon_r(t) = \frac{\epsilon_r(t_0) R_0^4}{R^4(t)} \tag{2}$$

where $\epsilon_r(t_0)$ is the present value of a radiation energy density that is a relic of an early hot era. The equivalent mass density $\epsilon_r(t)$ is

$$\rho_r(t) = \epsilon_r(t)/c^2 \tag{3}$$

and therefore also decreases as $R^{-4}(t)$. Therefore as we go backward in time the radiation density increases faster than the matter density, so that, however small ϵ_r is now, there must have been a time t_E when the densities of radiation and matter were equal. At such a time we have

$$\rho(t_E) = \left(\rho_0 R_0^3\right)/R^3(t_E) = \rho_r(t_E) = \left(\rho_r(t_0) R_0^4\right)/R^4(t_E)$$

that is,

$$\rho_r(t_0)/\rho_0 = R(t_E)/R_0 . \tag{4}$$

At times earlier than t_E, ρ_r was greater than ρ, and radiation was more dominant. Gamow therefore assumed that in the early epochs the dynamics of expansion were determined by radiant energy. The period

$0 < t < t_E$ is the **radiation-dominated era** of the history of the universe. The contents of the radiation-dominated universe are often referred to as the "primeval fireball", or as "ylem" (from the Greek "hyle" meaning "that on which form has yet to be imposed").

There is good reason to believe, as we shall see later, that the radiation and matter (in the form of relativistic particles and antiparticles) were in thermal equilibrium at the same temperature T, and that the radiation had a blackbody spectrum. As the universe expanded, the mean photon energy decreased as $R^{-1}(t)$ because of the redshift, so that the temperature T, which is proportional to this mean energy, should also have decreased as $R^{-1}(t)$. Since the radiation preserves its black body character (proof later), T continues to be proportional to $R^{-1}(t)$. Thus if T_0 is the present radiation temperature, then

$$T(t) = \frac{R_0}{R(t)} T_0 \tag{5}$$

which says that the temperature was very high at small values of R. We can carry the calculation one more step to get an explicit formula which relates radiation temperature T to time t. Since the radiation was in black body form with temperature T,

$$\epsilon_r = aT^4 \tag{6}$$

where "a" is the radiation constant (= 7.5×10^{-16} J m^{-3} K^{-4}). This means that in the early universe

$$T^0_0 = aT^4, \quad T^1_1 = T^2_2 = T^3_3 = -\frac{1}{3}aT^4 \ . \tag{7}$$

The space-curvature parameter k, from Eqs.(10-17) when applied to the present epoch, is given by

$$k = (2q - 1)H_0^2R_0^2/c^2 = (2q - 1)\dot{R}_0^2c^{-2} \ . \tag{8}$$

Since k will not affect the dynamics of the early universe significantly, we set it equal to zero. Thus, Eq.(10-8) becomes

$$\frac{\dot{R}^2}{R^2} = \frac{8\pi Ga}{3c^2}T^4 \ .$$

Substituting Eq.(5) into the above equation, we get

$$\dot{R} = A^2\left(\frac{8\pi Ga}{3c^2}\right)^{1/2}\frac{1}{R}; \qquad A = R_0 T_0$$

which can be easily integrated. Setting R = 0 at t = 0 we get

$$R = A(32\pi Ga/3c^2)^{1/4} t^{1/2} . \tag{9}$$

Eliminating R from Eqs.(9) and (5) we obtain the desired result

$$T = (3c^2/8\pi Ga)^{1/4} t^{-1/2} . \tag{10}$$

All the quantities inside the parentheses on the right-hand side are known physical quantities. Thus we obtain, after substituting numerical values for all the known physical quantities

$$T(K) = 1.52 \times 10^{10} t^{-1/2} (sec) \tag{11}$$

and we can see that about one second after the big bang the radiation temperature of the universe was on the order of 10^{10} K.

The idea of a hot big bang, as the above picture is called, depends on the basic assumption that the primeval radiation had a black body spectrum and that it preserved its black body character as the universe expanded. We also assumed that there is relic radiation present today. First of all, we need to address why we believe the primeval radiation had a black body spectrum.

Immediately following the big bang, the universe was so incredibly hot that all matter behaved like photons, since particles (whatever that may be!) move essentially at the velocity of light. Hence the initial state was a chaotic, gaseous inferno of high energy photons and elementary particles. We take up the story from the state when baryons (protons and neutrons), leptons (electrons, muons, neutrinos, and their antiparticles) and photons (particles of light) are already in existence. The particles would interact and collide, but these instances must have occupied very brief time spans, and so their effects on motions may be otherwise neglected. That is, these particles act as particles of an ideal gas. However, the collisions and scatterings of the particles would have helped to redistribute their energies and momenta. The radiation (photons) therefore stayed in thermal equilibrium with the matter at the same temperature and had a black body spectrum.

Our next task is to prove that the radiation should preserve its black body character. The proof is based on the conservation of photon number. The number of photons can only be significantly altered by interactions involving matter, but the number of elementary particles, as we shall see later, is far smaller ($\sim 10^{-9}$) than the number of photons — even though matter presently dominates energetically and gravitationally. Now the number dN(t) of photons in the frequency band ν and $\nu + d\nu$ in a volume V(t) of space at cosmic time t is given by Planck's law

$$dN(t) = \frac{8\pi\nu^2 V(t)\,d\nu}{c^3[\exp(h\nu/kT(t)) - 1]} \,. \tag{12}$$

As time proceeds, the number of photons in the volume remains the same, because of the conservation of photon number. This means that a co-moving observer would see as many photons crossing the (imaginary) boundary into this volume as leaving it. At a new time t′, the original group of photons has been red-shifted to a frequency

$$\nu' = \frac{\nu R(t)}{R(t')}, \qquad d\nu' = d\nu \frac{R(t)}{R(t')} \tag{13}$$

while the volume has expanded to

$$V(t') = V(t) \frac{R^3(t')}{R^3(t)} \,. \tag{14}$$

Therefore for this group of photons we have

$$dn(t') = dN(t) = \frac{\dfrac{8\pi}{c^3}\left(\dfrac{\nu' R(t')}{R(t)}\right)^2 V(t') \dfrac{R^3(t)}{R^3(t')} d\nu' \dfrac{R(t')}{R(t)}}{\exp(h\nu' R(t')/R(t)kT) - 1}$$

$$= \frac{8\pi V(t')\nu'^2 d\nu'}{c^3(\exp(h\nu'/kT(t')) - 1)} \,. \tag{15}$$

This looks just like a black body spectrum at a new temperature T(t′), where the new temperature T(t′) is

$$T(t') = [R(t)/R(t')]T(t) . \qquad (16)$$

Thus the radiation keeps its black body character, but it will appear cooler by a factor R(t)/R(t′). We should expect to see a low temperature relic radiation background. Gamow and his co-workers predicted in 1948 that this relic radiation should currently have a temperature of about 5 K. A black body radiation with temperature of this order is predominantly in the microwave form. Gamow's prediction was not taken seriously. However, relic radiation was indeed detected accidently by Penzias and Wilson in 1965.

Penzias and Wilson of the Bell Laboratories were using a 20 foot horn reflector radio telescope for both communications and radio astronomy. To have accurately calibrated results, they had to understand all sources of noise in their system. They found an unaccounted-for source of noise at a very low level. The noise seemed to be coming either from within their system or from every direction in the sky. After carefully analyzing their system, they were confident that the noise was coming from everywhere in the sky. They only had a measurement at one wavelength, so they could not confirm the shape of the spectrum. However, they found that the intensity corresponds to a black body at a temperature of about 3 K.

Astronomers and physicists have made many measurements of the intensity of this relic background radiation, at a variety of wavelengths. The most accurate measurements come from the Cosmic Background Explorer satellite (COBE), which was placed in orbit about the Earth in 1989. Data from COBE's spectrometer are shown in Figure 1, and demonstrate that this relic radiation has the spectrum of a black-body with a temperature of 2.735 K. (This relic radiation is now commonly called the 3-degree cosmic microwave background radiation). The little squares on the graph are COBE's measurements of the brightness of the cosmic microwave background plotted against wavelength. The peak of the curve occurs at a wavelength of 1.1 mm, in accordance with Wien's displacement law. An important feature of this cosmic microwave background is that its intensity is almost perfectly isotropic. We will return to this feature later.

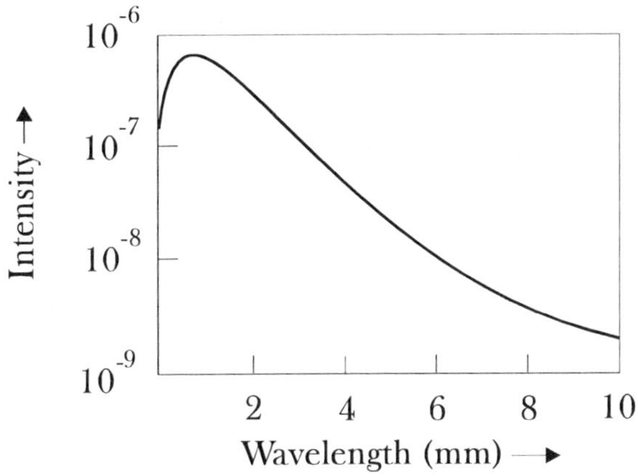

Figure 1: The spectrum of the cosmic background radiation (COBE).

The present value of the energy density of this cosmic microwave background is

$$\epsilon_r(t_0) = aT^4 = (7.5 \times 10^{-16}\,\mathrm{J\,m^{-3}\,K^{-4}}) \times (2.7\,\mathrm{K})^4 = 4 \times 10^{-14}\,\mathrm{J\,m^{-3}} \quad (17)$$

The equivalent mass density is

$$\rho_r(t_0) = 4.5 \times 10^{-31}\,\mathrm{kg\,m^{-3}}\ . \quad (17a)$$

This radiation mass density is only about a thousandth of the value for the matter density ρ_0, and so the universe is indeed matter dominated at the present time. The universe, of course, contains all other kinds of radiation; but this is not distributed with a black body spectrum and the total non-black body energy density has been estimated as less as $\epsilon_r(t_0)/100$.

Before we conclude this section, let us comment on the transition temperature and the baryon/photon ratio:

The transition temperature T_t

As we saw above, at times earlier than t_E the universe was dominated by radiation and the radiation and matter were in thermal equilibrium at the same temperature. At the time t_E, the universe entered the matter-dominated era and the radiation temperature was no longer equal to the matter temperature. We can make an estimate of the transition temperature T_t. From Eq.(4), if we take $\epsilon_r(t_0) = 4.5 \times 10^{-31}$ kg m^{-3} and $\rho_o = 3 \times 10^{-28}$ kg m^{-3}, we get

$$R(t_0)/R(t_E) \simeq 700$$

where we have used the lowest estimate of ρ_0. If "missing mass" exists, then $R(t_0)/R(t_E)$ will be greater than 700. From Eq.(5) and if we take the value 2.7 K for the present radiation temperature, we get the transition temperature T_t

$$T_t = 1900 \text{ K}.$$

Again, if there is "missing mass" T_t will be greater than this. Now this value of 1900 K is of the same order as 4000 K, above which a dilute gas of hydrogen would be almost completely ionized. Thus, as we shall see later, the transition period between the radiation- and matter-dominated eras, $t > t_E$, is also the recombination era during which the plasma of free electrons and protons condenses into neutral hydrogen which is much less opaque to radiation than plasma.

The baryon/photon ratio

As we saw above, the relic radiation of the hot big bang had a black body spectrum and it preserved this black body character as the universe expanded into the matter-dominated era and even to the present time. The proof is based on the conservation of photon numbers, and the photon number conservation is based on the assumption that the ratio of the elementary particle to photon is negligibly small ($\sim 10^{-9}$). To back up our assumption, let us make a simple calculation on the baryon/photon ratio. The energy density associated with black body radiation of temperature T is aT^4, and the mean energy per photon is $\sim kT$, so the number density of black body photons is $\sim 1.5 \times 10^{-9}$ for T = 2.7 K:

$$N_{ph} = aT^4/kT = aT^3/k = 1.07 \times 10^2 \text{ cm}^{-3},$$

$$a = 7.56 \times 10^{-15} \text{ erg cm}^{-3} \text{ k}^{-4}, \quad k = 1.38 \times 10^{-16} \text{ erg K}^{-1}.$$

The number density of baryons equals ρ_m/m_p, where m_p is the mass of the proton (= 1.66×10^{-24} g) and ρ_m is the mass density of the universe. If we take $\rho_m = \rho_0 = 3 \times 10^{-31}$ g cm^{-3} for the present epoch, we find the number density of baryons is $\sim 1.8 \times 10^{-7}$ cm^{-3}. Thus, the baryon/photon ratio is approximately equal to 10^{-9}:

$$N = 1.8 \times 10^{-7}/1.07 \times 10^{2} = 10^{-9}.$$

11-2 The creation of matter and photons

There are many misconceptions about the big bang. It was not at all like an exploding bomb. At the moment of the big bang, space and time are completely jumbled up in a state of infinite curvature like that at the center of a black hole. Thus, we should think of the big bang as an explosion of space at the beginning of time. We cannot use the existing laws of physics to tell us exactly what happened at the moment of big bang and what existed before the big bang. Without a clear background of space-time, concepts such as 'past', 'future', 'here' and 'now' cease to have meaning. A question naturally arises: where did all the matter and radiation in the universe come from? Recent intriguing theoretical research by physicists such as Zeldovich, Weinberg, and Parker suggest that the universe might have started as a perfect vacuum and that all the particles of the material world were created from the expansion of space-time. The detailed calculations are complex, but the basic ideas are not difficult to understand with the help of Heisenberg's uncertainty principle and the concept of quantum fluctuation. In Chapter 8, we have seen how the uncertainty principle and the concept of quantum fluctuation can help us to understand the Hawking effect. As we know, the vacuum of classical physics is an uninteresting empty state. However, quantum vacuum is a very interesting place, where energy exists although particles do not. Fluctuations of quantum vacuum can lead to the temporary formation of particle-antiparticle pairs. Normally, the mass Δm of the pair and its life-span Δt must satisfy the "Heisenberg-Einstein" relation:

$$\Delta m \times \Delta t \geq h/(2\pi c^2) \tag{18}$$

so that annihilation follows closely upon creation and the pairs are virtual (Figure 2a). During the big bang, however, space-time was expanding so fast that the particle and antiparticle may be pulled apart and gain real existence (Figure 2b). Were photons also created from the expansion of

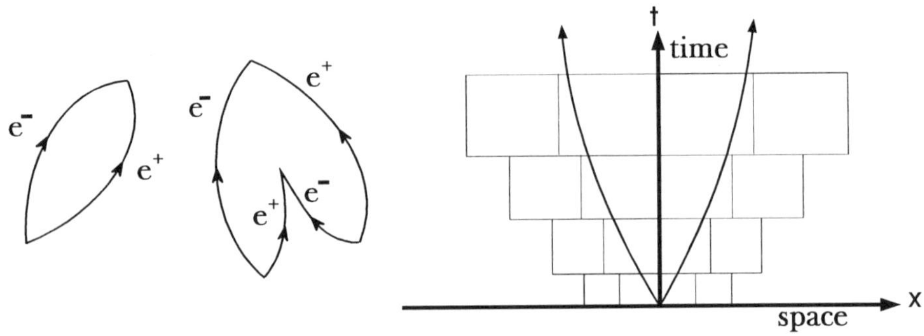

(a) Creation and annihilation (b) In expanding space-time

Figure 2: Pair production.

spacetime? The answer is no. Quantum mechanics forbids the direct creation of photons from the vacuum. Thus, particles with rest mass have to be created first. We now make an order-of-magnitude estimate for the particle creation from the expansion of space-time.

Suppose at the end of creation the particles are pulled apart by a distance d in a time interval Δt, and if the relative acceleration during the separation is g, then

$$d \sim g(\Delta t)^2 \,. \tag{19}$$

If gravity is the only force field in the early universe, we may write

$$g \sim (\ddot{R}/R)d \sim d/t^2 . \qquad (20)$$

Note that here t is the age of the universe. Substituting Eq.(20) into Eq.(19) shows $\Delta t \sim t$, i.e., the interval over which the particles can be pulled apart is comparable to the age of the universe. The particles move essentially at the speed of light, so we may write

$$d \sim ct .$$

Now, with $\Delta t \sim t$ and $d \sim ct$, Eq.(18) implies that, at the end of the process

$$(2m)c^2t = (2m)c(ct) \simeq (2m)cd \simeq h/2\pi$$

from which we have

$$d \sim h/mc$$

where, for order-of-magnitude estimate, we have dropped the factor 2 and π. Thus, the creation of particles of mass m occurs primarily when the cosmic time t equals the Compton time h/mc^2 ($t \sim d/c \sim h/mc^2$). At the end of the process, the pair is pulled apart by a distance d approximately equal to the Compton wavelength h/mc. Roughly, one particle would be created per Compton volume $d^3 \sim (h/mc)^3$.

11-3 A brief history of the early universe

For a very short time interval after the big bang, space-time did not behave the same as we understand it today. This short time interval, called the **Planck time** (t_p), is given by the expression

$$t_p = \sqrt{Gh/c^5} = 1.35 \times 10^{-43} \text{ s}$$

where G is the gravitational constant, h the Planck constant, and c the speed of light. From the big bang to the Planck time 10^{-43} s later, physics fails us. Now let us see why this breakdown occurs.

General relativity states that, near a mass m, events within distances shorter than the Schwarzschild radius, r_g, are hidden from the outside world, where

$$r_g = Gm/c^2 .$$

On the other hand, quantum-mechanical uncertainty limits our ability to locate the event horizon on a scale smaller than the Compton wavelength

$$\lambda_c = h/mc .$$

We cannot be sure that the singularity is inside the event horizon; there is a probability that the singularity spends some time outside the event horizon. Thus general relativity breaks down. If we equate the expressions for r_g and λ_c, we may solve for m to obtain the Planck mass m_p:

$$m_p = \sqrt{hc/G} = 5.46 \times 10^{-8} \text{ kg} . \qquad (21)$$

The length r_g or λ_c that corresponds to $m = m_p$ is called the Planck length λ_p

$$\lambda_p = \sqrt{Gh/c^3} = 4.05 \times 10^{-35} \text{ m} . \qquad (22)$$

The Planck time is the light-travel time across the Planck length:

$$t_p = \lambda_c/c = (Gh/c^5)^{1/2} = 1.35 \times 10^{-43} \text{ s} . \qquad (23)$$

Because general relativity breaks down for times earlier than the Planck time, no one knows how space, time, and matter behaved during the Planck time. A quantum theory of gravity is needed to describe properly the space-time under these extreme circumstances. No such theory exists yet. Nevertheless, Wheeler has speculated that space-time as we know today burst forth from a seething, foamlike, space-time mishmash during the Planck time. We can think of the big bang as an explosion of space at the beginning of time.

After the Planck time, space and time began to behave in the way we think of them today. We can safely apply the laws of physics to study the early universe. Temperature played a crucial, controlling role in the evolution of the early universe. Each period of time, after the Planck time, can be matched with a corresponding temperature via Eq.(11). Eq.(9)

gives the scale factor R as a function of time t. From Eqs.(9) and (11), or Eq.(5) alone, we see that temperature T is inversely proportional to the scale factor R.

As soon as the flood of particles and antiparticles appeared in the universe, collisions between them began to produce high-energy gamma ray photons which can be converted into particles and antiparticles again. An equilibrium was quickly established. For any particular particle of mass m there is an epoch at which the temperature is such that the energy of a typical photon is comparable to the rest-mass energy of the particle, that is, $kT \sim mc^2$. Prior to this epoch a collision between two photons can result in the creation of a particle and antiparticle pair. The thermal equilibrium between matter and radiation ensures that at these earlier epochs, when $kT > mc^2$, there are roughly as many of these particles as photons. When the temperature drops below mc^2/k, particle pairs can no longer be created, and so the particles and antiparticles rapidly annihilate each other. There are many different kinds of particles, each having its own threshold temperature. Threshold temperatures of several common types of particles, with their masses, are listed in Table 1.

Table 1: Threshold temperature of selected particles.

Particles	Rest energy (MeV)	Threshold temperature (10^9 K)
Neutrino ν	0.00001(?)	0.00001
Electron $e^{\pm}$	0.5110	5.930
Muon $\mu^{\pm}$	105.66	1,226.2
Pion $\pi^{\pm}$, π°	134.96	1,556.2
	139.57	1,619.7
Proton p	938.26	10,888
Neutron n	939.55	10,930

As the universe expanded adiabatically, the temperature dropped. When the universe was about 10^{-4} second old (that is, at time $t = 10^{-4}$ s after the big bang), the temperature of the universe fell below 10^{13} K (see Eq.11) which is cooler than the threshold temperatures of both protons and neutrons, so the annihilation of protons by antiprotons and of neutrons by antineutrons proceeded vigorously. This wholesale annihilation dramatically lowered the matter content (particles and antiparticles) of the universe, while simultaneously increasing the radiation (photon) content.

From Table 1 we see that for T < 1.2 x 10^{12} K the only particles that can be present with appreciable number densities in thermal equilibrium are $\mu^{\pm}$, $e^{\pm}$, ν_e, $\bar{\nu}_e$, ν_μ, $\bar{\nu}_\mu$, and γ. The γ's, $e^{\pm}$, and $\mu^{\pm}$ follow their respective distribution functions. (Remember that photons are bosons, while electrons and muons are fermions). The neutrinos require some attention. The neutrinos are absorbed, emitted, or scattered in weak interactions such as the following:

$$p + e^- \rightleftarrows n + \nu, \qquad p + \bar{\nu} \rightleftarrows n + e^+$$

$$e^- + \mu^+ \rightleftarrows \nu_e + \bar{\nu}_\mu, \qquad e^+ + \mu^- \rightleftarrows \bar{\nu}_e + \nu_\mu$$

$$\nu_e + \mu^e \rightleftarrows \nu_\mu + e^-, \qquad \bar{\nu}_e + \mu^+ \rightleftarrows \bar{\nu}_\mu + e^+$$

$$\nu_e + e^- \rightleftarrows \nu_e + e^-, \qquad \bar{\nu}_e + e^+ \rightleftarrows \bar{\nu}_e + e^+$$

$$\nu_\mu + \mu^+ \rightleftarrows \nu_e + e^+, \qquad \bar{\nu}_\mu + \mu^- \rightleftarrows \bar{\nu}_e + e^-$$

These are all examples of weak interactions. The early universe must have been populated with vast numbers of neutrinos and anti-neutrinos. As we will show later, when the temperature dropped below 10^{12} K, the neutrinos ceased to interact with the rest of matter and consequently dropped out of thermal equilibrium as temperatures fell appreciably below 10^{12} K. How far below? The original theory of weak interactions suggests that this temperature may be around 1.3 x 10^{11} K. We now summarize the events preceding the electron-positron annihilation.

When the universe was about 1 second old, its temperature fell below 6 x 10^9 K, electrons and positrons annihilated each other, further decreasing the matter content of the universe while raising its radiation content. Now we face a dilemma. If the symmetry between particles and antiparticles were perfect, then for every particle there should have been an antiparticle. Consequently, by the time the universe was 1 second old, every particle would have been annihilated by an antiparticle, leaving no matter at all in the universe. Evidently, this did not happen. The planets, stars, and human bodies are made of matter. Thus, there must have been an excess of matter over antimatter immediately after the big bang. Some physicists suggest that symmetry was broken during the earliest moments of the universe. As a result, the number of particles which emerged from the big bang was slightly in excess over anti-particles. By how much? There are roughly a billion photons today in the microwave background for each proton and neutron in the universe. Consequently, for every billion antiprotons, a billion plus one ordinary protons were created; and for every billion positrons, a billion plus one electrons were created.

Table 2:

Time (s) since big bang	Temperature (K)	Events
$< 10^{-4}$	$> 10^{12}$	Baryons, mesons, leptons and photons in thermal equilibrium
10^{-4} - 10^{-2}	10^{12} - 10^{11}	$\mu^{\pm}$ begin to annihilate and neutrinos begin to decouple from the rest of matter
10^{-2} - 1	10^{11} - 10^{10}	Neutrinos decouple completely. $e^{\pm}$ pairs still relativistic.

The number of protons was kept approximately equal to the number of neutrons through the reversible reactions

$$p + e^- \rightleftarrows n + \nu, \quad p + \bar{\nu} \rightleftarrows n + e^+ .$$

By the time at which neutrinos and antineutrinos ceased colliding with protons and neutrons in significant numbers, the natural tendency for neutrons to decay into protons took over, and the number of neutrons consequently declined. Before all the neutrons could decay into protons,

protons and neutrons combined to form H^2, H^3, He^3, and He^4 through sequences of 2-body reactions like

$$n + p \rightleftarrows H^2 + \gamma, \qquad H^2 + H^2 \rightleftarrows He^3 + n$$

$$He^3 + n \rightleftarrows H^3 + p, \qquad H^3 + H^2 \rightleftarrows He^4 + n \ .$$

The crucial step is the formation of deuteron (H^2), which, because of the low binding energy of the proton and neutron (2.2 MeV), is destroyed by collision as soon as it is made if the temperature is higher than about 10^9 K. Because deuterons could not survive, the helium nucleus could not be created. This block to helium creation is called the deuterium bottleneck. When the universe was about 3 minutes old ($T \sim 10^9$ K), the photons became so redshifted that they could no longer break up the deuterons. Because deuterons could now survive, helium nucleus was rapidly produced. The helium nucleus is very stable, and so by the end of the first three minutes after the big bang, 25% of the mass of the universe was tied up in He^4. The He^3 makes up only 10^{-4} to 10^{-5} of the total mass. The absence of stable nuclei with mass numbers $A = 5$ and $A = 8$ means that very little production of nuclei heavier than the helium nucleus could take place. We could say that big bang nucleosynthesis was effectively ended with the formation of He^4 nucleus. The universe was dominated by radiation and optically thick (that is, matter and radiation was strongly coupled).

Following big bang nucleosynthesis, the expansion of the universe continued with very little change until a time ($t > 10^5$ yr) when the temperature dropped to such a low values ($T < 4{,}000$ K) that the electrons and protons combined to form neutral hydrogen atoms. The matter therefore became transparent to radiation, and what we now observe as the 3 K microwave background radiation was released. The universe thereby entered into matter-dominated era.

When matter became transparent to radiation, gravity came into its own, dominating interactions between largely neutral masses of material making up the bulk of matter in the universe. Gravity caused galaxies, stars and planets to form. It is outside our syllabus to pursue the question of galaxy and star formation. Therefore we refer interested readers to special texts on galaxy and star formation.

We end this section by referring back to the primordial helium. A question which arises naturally is: how do know how much of the helium in the universe is primordial? We can get the answer by measuring the helium content of young and old stars. The old stars were formed when the galaxy was young, before the interstellar medium had been enriched by helium formed in stellar interiors. The young stars have been formed from an interstellar medium which had been enriched by helium formed in stellar interiors, plus primordial helium, if any. Therefore the comparison of the helium content in the two groups of stars tells how much helium is primordial.

The helium content of young stars can be determined most directly from the intensities of the helium absorption lines in their spectra. These lines are formed in the atmosphere of the star, and their intensity only gives the amount of helium in the star's outermost layer; however, in a young, unevolved star, the helium is dispersed uniformly, and the amount in the atmosphere is an accurate indicator of the amount in the entire star. Only the hot O and B type stars can be used, because helium lines only appear with significant intensity in the spectra of these stars. The spectroscopic studies show that a fraction 0.27 ± 0.05 of the mass of young stars consists of helium.

The O and B type stars are massive and last only a short time, not more than 100 million years. Therefore for old stars, we cannot use spectroscopic methods to determine the helium content. Instead, we can use H-R diagrams. We can determine the age of a globular cluster by fitting theoretical H-R diagrams computed for clusters of various ages to the observed H-R diagram. Computations on stellar structure show that the luminosity of a star, and, therefore, its position on the H-R diagram, depend on its chemical composition. In particular, the luminosity is strongly dependent on the helium content of the star. By matching the observed and theoretical H-R diagram for a globular cluster with care, it is possible to determine the helium content of the cluster as well as its age. The result is an age of roughly 10 billion years, and a helium content of 0.30 ± 0.05.

Thus, from the measurements we found that the helium content of the oldest stars in our galaxy is the same as the helium content of the youngest stars, within the limits of experimental uncertainties. In other words, nearly all the helium in our galaxy is primordial, and little has been added by nucleosynthesis in stellar interiors. This is a circumstantial evidence in favor of the big bang.

11-4 Problems with the standard big bang model

We have so far glossed over the drawbacks of the standard big bang model. We address two of them here: the flatness problem and the horizon problem.

(A) The flatness problem

The flatness problem was first pointed out in 1979 by R. H. Dicke and P. J. E. Peebles, and later highlighted by A. Guth. To appreciate it, let us recall Eq.(10-18) which we rewrite as

$$\frac{\dot{R}^2}{R^2} + \frac{kc^2}{R^2} = \frac{8\pi G\rho_0}{3}$$

where R(t) is the scale factor, $\dot{R}$ its rate of change with time, ρ_0 the density (matter and energy) of our universe, and the curvature parameter k can take the values 1, 0, -1. We next introduce the density parameter Ω:

$$\Omega = \frac{\rho_0}{\rho_c} = \frac{8\pi G\rho_0}{3H^2} \tag{24}$$

where H is the Hubble constant. The Friedmann dynamical equation becomes

$$\Omega = 1 + \frac{kc^2}{H^2R^2} = 1 + \frac{kc^2}{\dot{R}^2} \tag{25}$$

If $k = 0$, we have $\Omega = 1$. For $k \neq 0$, as we approach closer and closer to the big bang epoch, R(t) increases, and the second term on the right-hand side of the above equation becomes smaller and smaller, accordingly the density parameter approaches 1. In other words, at its beginning, the density of the universe must have been at or near the critical density point, even if it does not appear so today.

Current observations indicate that the universe today weighs in at about 15% of the critical density ρ_c. Instead of trying to find all of the matter in the universe, we can look for its gravitational effects, namely, by examining the deceleration parameter q_0 which is directly related to matter in the universe. The case of $q_0 = 1/2$ corresponds to a universe which just barely manages to expand forever. If $q_0 < 1/2$, the universe is unbounded and will continue to expand forever. Such a universe contains matter at less than the critical density. If $q_0 > 1/2$, the universe is closed and has matter with a density greater than ρ_c. The corresponding observations are very difficult to make. Figure 3 shows data obtained by Allan Sandage.

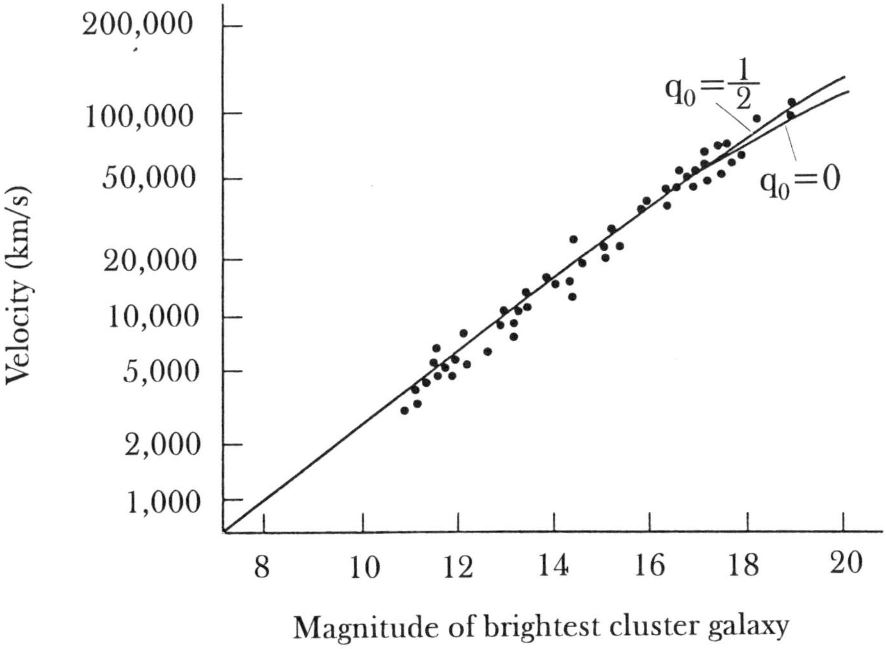

Figure 3: The Hubble diagram.

The data points are scattered about the line $q_0 = 1/2$. This scatter prevents us from determining conclusively whether the universe is open or closed, but, because the data lie close to the line $q_0 = 1/2$, the universe is very nearly flat and contains matter at or near the critical density ρ_c.

The problem is that given all of the infinite possible masses that our universe could have, why does it have a mass so close to the critical value? Because $\rho_0 = \rho_c$ means that our universe is flat, this puzzle is called the flatness problem.

The earliest understandable moment in the universe was the Planck time, which is about 10^{-43} second after the big bang. Before the Planck time, the universe was so dense and particles were interacting so violently that no known theory can describe what was happening then. What, then, could have happened immediately after the Planck time to ensure that $\Omega = 1$ (or $\rho_0 = \rho_c$) to a very high degree of accuracy?

(B) The horizon problem

A second enigma, closely related to the flatness problem, is called the horizon problem and concerns the isotropy of the 3-K cosmic microwave background radiation. As we saw in Chapter 10, the 3-K background radiation deviates from complete uniformity by only 1 part in 100,000! This tells us that the universe was extremely isotropic at the time of decoupling and the 3-K background radiation was emitted from regions with a common temperature. This means that all parts of the universe must have been in communication (or interaction) with each other before that time. How did this happen? Therein lies a problem.

The horizon problem was pointed out by Wolfgang Rindler in 1956. To understand the horizon problem we need to introduce a new quantity, namely the horizon distance D. D is simply the age of the universe (t_0) multiplied by the speed of light (c), or

$$D = c \times t_0 \,.$$

No information can be received from a distance farther than D, so D is called the horizon distance (or radius of communication). As the universe expands, D increases faster than the scale factor R. Conversely, as we go back in time, D decreases faster than R.

The number of photons inside the universe does not change as the universe expands or contracts. Estimates place this number (of photons) in the universe at about 10^{87}. Let us assume that all 10^{87} photons are observable today, so the horizon distance D must equal the scale factor R. If we imagine ourselves going back in time toward the big bang, D decreases faster than R, so we see a smaller and smaller fraction of the 10^{87} photons in the universe. A simple calculation shows that at one second after the big bang, D was only about 10^{-9}R and only 10^{60} photons could be observed. At 10^{-35} sec, D was about 10^{-27}R and roughly one million photons were within the horizon. At 10^{-43} sec, essentially no photons were within the horizon.

Since a particle (including a photon) cannot exchange information with another particle outside its horizon, the above thought experiment clearly show that virtually all particles were incommunicado in the early universe, and they had no way to smooth out any irregularities. That is the horizon problem.

If you have any reservation or doubt about the above thought experiment, try the following approach. Consider two nearby points in the sky as shown in Figure 4, θ_0 ($<< 1$ rad) is the current angular separation, and x_0 is the linear separation. If the points are a distance d from us, then

$$x_0 = \theta_0 d \ .$$

If we are just seeing light from these points, it must have been travelling for $t_0 - t_d = t_0$, where t_0 is the current age of the universe, and t_d the age of the universe at decoupling which occurred very early ($\sim 10^5$ years) in the history of the universe. Thus

$$d = (t_0 - t_d)d = t_0 d$$

and so

$$x_0 = \theta_0 c t_0 \ .$$

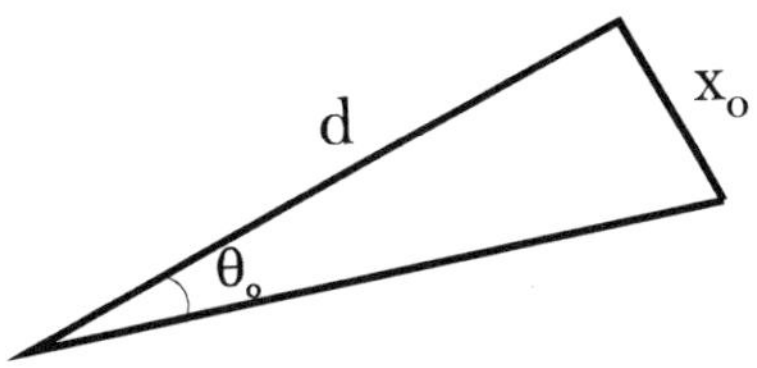

Figure 4: Causality and the isotropy of the background radiation.

If we take our reference epoch to be now, so that $R(t_0) = 1$, and let R be the scale factor at the time of decoupling, then the linear separation x of these two pints at the time of decoupling is R times their current linear separation, or

$$x = Rx_0 = R\theta_0 ct_0$$

The time for light (signal) to travel the distance x is x/c, or

$$t_x = x/c = R\theta_0 t_0 \ .$$

For these two points to interact or to communicate, t_x must be less than or equal to t_d. The farthest the two points can be is when $t_x = t_d$, so

$$t_d = R\theta_o t_o$$

or

$$\theta_0 = t_d/(Rt_0) \ .$$

Now $t_0 = 10^{10}$ yr and $t_d = 10^5$ yr, and if we know the scale factor R at the decoupling, we can calculate θ_0. We know that $T(t) \propto 1/R(t)$, and the background temperature (3 K) was 3000 K at decoupling. So R at decoupling must have been 10^{-3}. Accordingly

$$\theta_0 = (10^5 \text{ yr})/(10^{-3})(10^{10}\text{yr}) = 10^{-2} \text{ rad}$$

which is approximately 1°.

The above simple calculation tells us that points that are more that one degree apart cannot be causally connected. That is, the 3-K background radiation should not appear smooth on scales larger than one degree.

In the early 1980s, Alan Guth at Stanford University offered an inflationary theory as a solution to the problem of flatness of the universe and the isotropy of the 3 K background radiation (the horizon problem). Alan Guth's theory owes its origin to the curious effects of phase transition, which is believed to have occurred in the very early epochs of the universe when the breakdown of the so-called grand unification symmetry took place spontaneously. Before we discuss Alan Guth's inflationary theory, it is worthwhile for us to become familiar with these jargon of quantum field theory and high-energy particle physics.

11-5 Phase transition, spontaneous symmetry-breaking, and GUTS

In physics a phase transition is the transformation from one physical state to another. Water is a good example to illustrate the fundamental concepts of phase transition and spontaneous symmetry-breaking. We all know that water can exist in three physical states: vapor (the gaseous state), the liquid state, and ice (the solid state). Therefore water has three possible phase transitions. The phase transition from liquid to solid gives the best illustration of spontaneous symmetry-breaking. The liquid and solid states of water have two essential differences:

1. The liquid water is completely isotropic: there is no preferred direction so the properties of liquid water are the same in every direction. In the crystallized ice the symmetry is broken. Along the lines of the crystal lattice the physical properties differ from those observed in other directions. Thus, whenever there is a transition between the two states, the symmetry is broken in the physical theory that describes the properties of water.

2. The crystallized ice has less energy than the liquid state. Thus, when we want to melt ice we have to heat it. Conversely, as water turns into ice we have to remove heat energy.

Another familiar phase transition and spontaneous symmetry-breaking occurs in a ferromagnet, a magnetized iron bar. Every atom of iron can be thought of as a tiny magnet with a north and a south pole. When a large number of these tiny magnets are lined up with the north poles in the same direction, the iron is said to have a net magnetization. Now consider a piece iron that has been heated to a high temperature, and consider what the energy of the system looks like. Because of the heating, the iron has no net magnetization. If we want to give the system some magnetization, it will be necessarily to add energy to create the alignments. Thus, heated iron with a net magnetization must be at a higher total energy than heated iron which has lost all magnetization. If we plot energy of the system against magnetization, we will get a graph like Figure 5a. The point zero magnetization is a minimum — it corresponds to the ground state or the vacuum state.

Now, suppose we lower the energy of the iron by allowing it to cool. At the Curie temperature (~ 770°C) a phase transition occurs and the energy curve suddenly transfers to a more irregular shape like the one shown in Figure 5b. The lowest energy state does not correspond to zero magnetization but to some value different from zero. The process behind the changing of the potential curve is simple: at the Curie temperature the interaction between atoms is such that they tend to align themselves to produce a magnetic field. No outside intervention is needed to achieve this alignment; it occurs spontaneously when the temperature is right. There is no overall change of energy involved in the process, just the establishing of order among the atoms in the system, with an attendant redistribution of energy between atoms and the magnetic field.

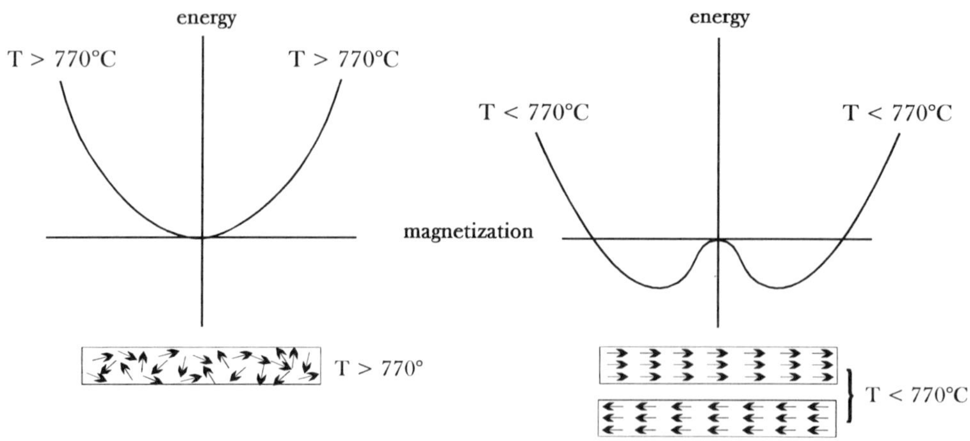

(a) Below Curie temperature (b) above Curie temperature

Figure 5: A schematic curve of potential energy.

If we define the true ground state of iron to be the one associated with zero magnetization, then the lowest energy state in Figure 5b is a false ground state (or a false vacuum).

The ferromagnet analogy also illustrates the concept of spontaneous symmetry-breaking. Above the Curie temperature, there is no net magnetization. Every direction is the same in the sense that no matter in which direction we look, we see no net magnetization. We say that above Curie temperature there is a high degree of symmetry in the system. Below the Curie temperature there is a preferred direction in space, the direction defined by the N and S poles of the magnet. Therefore, at the Curie temperature the symmetry is broken spontaneously (without interference from outside agencies) and iron falls into a false ground state, a state of lowest energy but not zero magnetization.

The forces (interactions) of nature

There are four forces in nature which hold matter together through the various subtleties of attraction and repulsion. The forces of gravity and electromagnetism are familiar in daily life. The strong force (responsible for binding the parts of a nucleus together) and the weak force (governing radioactive decay) operate at the subatomic level and are not immediately available to our senses. However, in the very early epochs of the big bang, the universe was packed so closely together that these two forces were significant. Over the decades, physicists have pondered how these forces of nature are tied together and work in concert. A. Einstein was among them. He spend over 20 years (until his death in 1955) trying to unite electromagnetism with gravity in his unified field theory. Although never successful, he set the stage for later work that was able to take into account both strong and weak forces.

Today, the force between two particles is described in terms of an exchange between them involving one or more other particles which serve as intermediaries and they are called the carrier (or vector) particles of the interaction. The masses of the carrier particles determine the ranges of the different interactions. For example, the quantum theory of electromagnetism, which is called quantum electrodynamics, describes the interaction between two charged particles with an exchange of photons.

These photons are the carrier particles of an electromagnetic interaction in which the electrons or protons are the actors. The mass of the photon is zero and so the virtual vector particle has an infinite domain.

There were only four known particles at the beginning of the 1930, the electron, the proton, the neutron, and the photon. Today, we have hundreds of particles. The "material" particles are all fermions (particles with half-integral spins), and the force carriers are bosons (particles with integer spins). Strongly interacting particles are known as hadrons, while those that do not feel the strong force are leptons. All leptons are fermions. The hadrons that are also fermions are called baryons. Protons and neutrons are baryons. The bosons that carry the strong forces between particles are known as mesons. The point is a meson and comes in three varieties: π^+, π^-, and π^0.

To bring some order to the proliferation of particles, the "quark" concept was introduced in the middle by Murray Gellmann in 1964. He proposed that protons, neutrons and a hundred or so "elementary" particles are made up of two quarks: u (up) quarks and d (down) quarks; the u quark has charge 2e/3 and the d quark -e/3. Both u and d quarks are fermions. Furthermore, quarks are endowed with a new sort of charge, called color charge. The color charge of each quark can assume one of three values: R(red), B(blue) or G(green). For every quark there is an antiquark with the opposite properties associated with antimatter. In this theory, baryons are made of three quarks, and their carrier particles (mesons) are made up of a quark and an antiquark (not necessarily of the same type as the quark). For example, the proton is uud and the neutron udd; the neutral pion is u and anti-u, and the positive pion is u, anti-d.

At the present time, the hundreds of known particles can be accounted for by means of three families of quarks: the first with u and d, the second with s (strange) and c (charm), and the third with b (beauty or bottom) and t (truth or top). The existence of five of these six varieties of quarks, or "flavors", has been confirmed experimentally, and the existence of the t quark has been inferred at CERN. Leptons are not made of quarks.

We are now ready to embark on the question of how to unify the forces (interactions) of nature. To keep things simple we will concentrate on fermions in the first family, namely the electron, the electron neutrino, and six quarks (u and d each in three colors).

The strong interaction can be described in terms of the quark-quark interactions. The quark-quark interactions take place through gluons, the virtual carrier particles of the interaction. We can summarize this interaction in a table. For u quarks for example, it looks like this:

	uR	uG	uB
uR	g(R-R)	g(R-G)	g(R-B)
uG	g(G-R)	g(G-G)	g(G-B)
uB	g(B-R)	g(B-G)	g(B-B)

From this table we can see that a B quark can turn into an R quark by emitting a (B-R) gluon.

Gluons, like photons, have zero mass and travel with the speed of light. The range of the strong interaction is, consequently, infinite. Nonetheless, nucleons themselves interact over minute distances less than 10^{-15} m. We must consider them as complex systems. Their behavior is similar to that of molecules: molecules also interact over a short distance even though the electromagnetic interaction has an infinite range.

Photons do not have an electric charge, but gluons have the color charges, represented by suitable combinations of the colors R, G, or B. For example, an R quark emitting a R-G gluon becomes a G quark. (This is a complicated effect, and that is why chromodynamics is more complex than quantum electrodynamics.)

In electromagnetic interaction, the active particles (electrons, protons and so forth) and their carriers (photons) can exist in their free state. However, neither the quark nor the gluon are ever seen as free particles. Is this a fatal blow to the theory? The answer is no. One possible way out comes from the curious phenomenon of the polarization of the vacuum which is allowed by Heisenberg's uncertainty principle. The importance of vacuum polarization (or vacuum fluctuations) in electromagnetic processes has long been experimentally confirmed (See Section 6 of Chapter 8 on Lamb shift of the spectral lines of the hydrogen atom.) The vacuum polarization comes into play for the quark in a little more complex way. The quark, denoted by q, is surrounded by a sea of virtual gluons as well as pairs of quarks and antiquarks. These virtual gluons, unlike photons which have no electric charge, have a color charge; rather than shielding q, they reinforce and extend q's charge. The net effect of the vacuum polarization is to amplify the intensity of the strong interaction at a distance and to decrease it locally. At very short distances the intensity of the strong interaction asymptotically approaches zero, and quarks become independent of each other at very close distance. However, if we try to separate them from each other, we have to struggle against an interaction which becomes stronger and stronger, and thereby endows the quarks with more and more energy. A final stage will be reached at which this energy, rather than being used to separate the quarks, instead gets used in the creation of quarks and antiquark pairs, which in turn form new nucleons and pions. Thus, it is impossible to observe free quarks.

The weak interaction works among particles that carry the weak charge and has a very short range, so it is mediated by heavy particles. According to the theory, there are three such heavy particles, all having spin 1; the two charged members of the family are called the W^+ and W^-, respectively, while the neutral member is called the Z^0. We have two tables:

	e	ν
e	W^0	W^-
ν	W^+	W^0

	u	d
u	W^0	W^+
d	W^-	W^0

so a u quark can turn into a d quark by emitting a W^+ boson. Figure 6 describes the decay of the neutron $n \rightarrow p^+ + e^- + \bar{\nu}$:

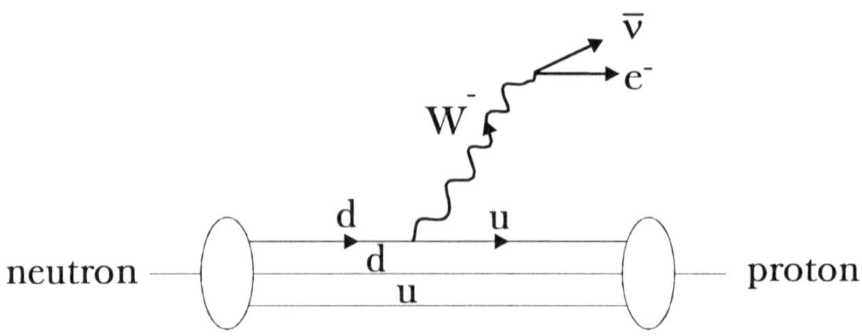

Figure 6: Decay of a neutron.

In 1967 Weinberg, Salam and Ward successfully united the electromagnetic and weak interactions. The combined interaction is given the name electroweak interaction. The unification was accomplished through the introduction of a new particle, the Higgs boson, which is massive but has no spin.

The vacuum is without any Higgs bosons. In order that the vacuum be deprived of Higgs bosons, it has to be provided with enough energy to quench their intensity. Accordingly, the vacuum, without any Higgs bosons, is rich with latent energy waiting to be liberated. This "vacuum dynamite" will play a crucially important role in the inflationary scenario.

A further property of Higgs particles is to increase the mass of the W particle associated with the weak interaction. The W^0 boson thus modified is able to interact with a photon and turn into a new boson Z^0, which has a mass different from W^+ and W^-. The electroweak interaction table looks like this:

	e	ν
e	γ; Z^0	W^-
ν	W^+	Z^0

The electroweak theory was confirmed in 1983 at CERN with the discovery of the W^+, W^- and Z^0 bosons, their masses tied exactly with the theory: 90 times the proton mass for W^+ and W^- bosons and 100 times for the Z^0 bosons.

The next stage of unification is to tie together the electroweak interaction with the strong interaction. Such a theory is referred to as **grand unified theory** (or **GUT**). Glashow and Georgi of Harvard University proposed a first grand unified theory in the 1970s. Since then different versions of grand unified theory have been proposed, all highly speculative. Although in specific details one GUT may differ from another, almost all of them involve gauge theories with a mediating role played by the Higgs particles. Let us combine two of the tables we have looked at: the one for the u quarks, and the one for the electroweak interaction. If we align the diagonals of these two tables we get a table with five rows and five columns which is partially filled:

	e	ν	uR	uG	uB
e	γ; Z^0	W^-		X	
ν	W^+	Z^0			
uR			g(R-B)	g(R-G)	g(R-B)
uG		X	g(G-R)	g(G-G)	g(G-B)
uB			g(B-R)	g(B-G)	g(b_b)

The top left section is filled with intermediate bosons (W^+, W^-, and Z^0) and the photon γ. At bottom right we only show the gluons associated with the strong interaction. To these we should add the intermediate bosons and photons that are also coupled to quarks. Two empty sections are marked X. The principle of GUT supposes that there are new bosons to fill the gaps. These new particles (12 in the simplest case) carry electric charge, weak charge, and color, and they allow interactions between leptons and quarks.

The electric charge of these hypothetical X bosons has fractional value (-4e/3, -e/3, +e/3, or 4e/3), and they are also extraordinarily massive, some 10^{15} times the mass of the proton. With this huge mass, the distance over which they exert an appreciable force is extremely small, 10^{-29} cm (given by the Compton wavelength). Such tiny yet extraordinarily massive X bosons are out of reach of present techniques. Many particle physicists have turned their attention to hot big bang where the available energy in the earliest epoch was large enough to create X particles and these should have left some observable traces, such as evidence for the inflationary phase of the universe.

It is clear that although the three interactions have all the appearances of being symmetrical and unified, they do not manifest themselves on an equal basis under all circumstances. This is due to the large differences between the masses of the vector particles — the photon, the intermediate boson (W, Z boson), and the X boson: respectively they are 0, 100 and 10^{15} times the mass of the proton. If particles were colliding at energies greater than 10^{14} GeV (equivalent to a temperature of 10^{27} K, E~kT), the strong, weak and electromagnetic interactions would be indistinguishable from each other. As the energy of the colliding particles fell below 10^{14}

GeV, the X bosons could no longer be created spontaneous by from the ambient energy, and the strong nuclear force is no longer unified with the electromagnetic and weak interactions. That is, at energy 10^{14} GeV (or at temperature 10^{27}K), there was a spontaneous breaking of symmetry. This situation existed in the very early universe. The critical temperature 10^{27} K corresponds to 10^{-35} second. As the universe expended further, at temperature 10^{15} K intermediate bosons were toppled also. This critical temperature corresponds to 10^{-11} second. The spontaneous breaking of symmetry at the two critical temperatures plays a crucially important role in releasing the latent energy of the vacuum as implied by the Higgs particles.

11-6 The inflationary scenario

The inflationary scenario was originally developed by Alan Guth. Important improvements were later added by A. Albrecht, P. Steinhardt, and A. Linde. The latent energy of the vacuum provides the energy in the inflationary phase of the early universe, and a phase transition releases that energy.

The vacuum in GUT is a very special one, its energy density (or its energy levels) depending on the density of Higgs bosons. Essentially there are two distinct states, such as we explored with water or a ferromagnet.

1. A high energy state, where the Higgs field has a value of zero and the density of Higgs bosons is zero. The vacuum is symmetric. This corresponds to liquid water, or a ferromagnet above the Curie temperature.

2. A state of much lower energy, where there are significant numbers of Higgs bosons and the Higgs field acquires a nonzero value. The vacuum is not symmetric. This energy state only exists below a critical temperature 10^{27} K.

What this means is that below the critical temperature 10^{27} K the vacuum can make a phase transition from a symmetrical state with no Higgs bosons, to a state of broken symmetry with some Higgs bosons at a much lower energy level. It is unclear how the symmetry broke down; scenarios reveal that it may have just happened spontaneously. The feature of interest to us is that the amount of energy released by the phase change is enormous, because there is a large energy difference between the high energy and low energy states. This transition energy is, therefore, the "dynamite" for inflationary. This critical temperature 10^{27} K corresponds to 10^{-35} second after the big bang.

At time before 10^{-35} second, or at temperature higher than the critical temperature 10^{-27} K, the highly condensed and extremely hot universe expands and cools, like general relativity and the equation of state for relativistic universe say it should. The strong and electroweak forces were unified and the Higgs field had a value of zero, in accordance with grand unification theory. At the moment 10^{-35} second, the temperature falls to 10^{27} K, and the symmetry is going to be broken. That is, the vacuum, which was symmetric, gets turned into a vacuum with broken symmetry via the phase transition. One of the consequences of this is that the Higgs field acquires a nonzero value. This phase transition may not take place instantaneously. This may be compared with the phase change from steam to water. Steam can be supercooled below the boiling point of water so that even at this lower temperature it remains in its gaseous state. But any small disturbance will immediately make it turn to water, which suddenly releases all the latent heat in one go.

Something similar to this supercooling process may happen to the universe. There may be a delay as the vacuum turned from a symmetric state to a state of broken symmetry. We may term the vacuum with broken symmetry as a false vacuum, because it retains the high energy associated with symmetric vacuum, and the value of the Higgs field remains at zero. We may envisage each unit volume of space as trapped in a false vacuum state of the Higgs field until it tunnels through the energy barrier and settles into the true vacuum (Figure 7). Although this dynamite (the tripled energy) does not explode, since the phase transition is not yet started, it does drastically change the behavior of the expansion by speeding it up tremendously compared to the normal big bang expansion. This acceleration is the secret of the inflationary theory. The concept, due to Alan Guth and A. Linde, is very hard to grasp. Let us sketch it below.

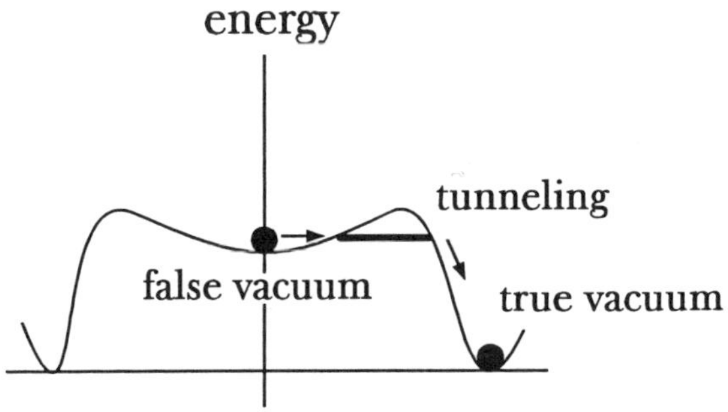

Figure 7: The universe is trapped in a false vacuum.

What happens is this: each unit volume of space is trapped in a false vacuum and contains an enormous amount of energy, and this remains so as space evolves with expansion. Thus as space expands, more and more energy is introduced into the universe as the total volume of the vacuum increases. As an analogy, this may be compared with a block of rubber: the more the rubber is stretched the more energy it has, because the elastic fibers store energy. The false vacuum behaves like a gas with "negative" pressure. The negative pressure would not produce mechanical forces (which arise from differences in pressure) within the false vacuum. But there would be gravitational effects. According to general relativity, the gravitational force is proportional to the energy density plus three times the pressure. For the false vacuum the contribution made by the pressure would overwhelm the energy-density contribution and would have the opposite sign. That is, the negative pressure leads to a "repulsive" gravitational force and introduces an extra term, the cosmological constant, into Einstein's equations. It acts on the expansion like a cosmic repulsion.

This cosmic repulsion, which is always of the same value and is inexhaustible, has the effect of stimulating the expansion of the universe at an explosive rate, exponentially in fact (we will comment on this exponential expansion later). The result is that space triples its size every 10^{-34} second after the epoch 10^{-35} second. This exponential expansion is termed "inflation" and would last until the epoch 10^{-32} second (as the theory suggests), when the transition to the broken symmetry phase was completed. Since there are 100 units of 10^{-34} second to use up before the elapsed time is 10^{-32} second, the tripling process takes place around 100 times: 3 x 3 x 3 ... = 10^{50} times altogether. During the phase transition, the latent heat was released, resulting in a tremendous amount of particle production. The region was reheated to a temperature of almost 10^{27} K. From this point on, the region would continue to expand and cool at the rate described by the standard big-bang model. The scenario is sketched in Figure 8. Note that the curves are not drawn to scale, nor are any specific numbers shown. These depend on the details of how the phase transition occurs.

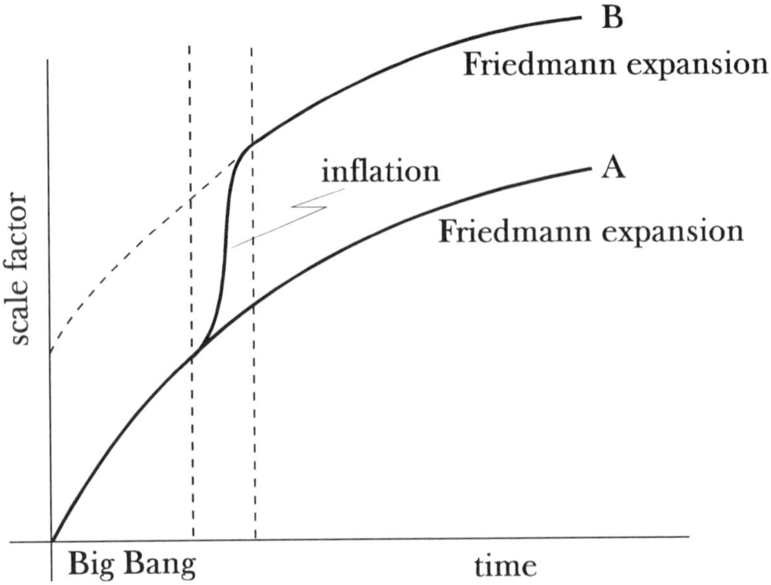

Figure 8: Plot of expansion showing inflation.

We stated earlier without proof that the expansion of the supercooled material of the universe is exponential. This can be determined by Einstein's field equations, which now take the following forms:

$$\frac{\dot{R} + kc^2}{R^2} = \frac{8\pi G}{3c^2}(u_r + u_v) \tag{26}$$

and

$$2\frac{\ddot{R}}{R} + \frac{\dot{R} + kc^2}{R^2} = -\frac{8\pi G}{c^2}(p_r + p_v) \tag{27}$$

where u_r and p_r are the energy density and pressure of the relativistic particles, with $u_r = 3p_r > 0$, and u_v and p_v are the energy density and pressure of the false vacuum, with $p_r = - u_r < 0$. The quantities u_r and p_r decrease with the expansion of the universe as $1/R^4$, but p_v and u_v stay constant as long as the false vacuum is maintained. This can be seen from Eqs.(26) and (27). We first differentiate Eq.(27) with respect to t, and then express the result as a linear combination of Eqs.(26) and (27). After a little algebraic manipulation we get

$$\frac{d}{dR}(u_v R^3) + 3p_v R^2 = 0 \ . \qquad (28)$$

Substituting the relation $u_v = - p_v$, we get u_v = constant. Furthermore, u_r and p_r tend to dominate the behavior of the solution of the equations, and also dominate the curvature term, k/R^2. Under these conditions, Eq.(28) takes the simple form

$$\frac{\dot{R}^2}{R^2} = a^2, \qquad \text{with} \qquad a^2 = \frac{8\pi G}{3c^2} u_v$$

or

$$dR/R = a\ dt$$

from which we can see easily that R has the exponential behavior described earlier.

How successful is the inflationary scenario in removing the flatness problem and the horizon problem? The horizon problem is solved in a straightforward way. According to the inflationary theory the observed universe evolves from a region that is 10^{50} times smaller in diameter than the corresponding region in the standard big bang model. Before inflation begins the region is much smaller than the horizon distance, and so it has time for every part of the universe to interact with every other part. Thermal equilibrium was therefore established before the inflation started, and the sources of the 3-K microwave background radiation arriving today from all directions in the sky were once at a common temperature.

The flatness problem is also evaded in a simple and natural way. Recall that the flatness problem arose from the comparison of the dynamic term $\dot{R}^2/R^2$ with the curvature term kc^2/R^2 in Einstein's equations. Now R~exp(at), the dynamic term stays constant, but the curvature term declines exponentially. Immediately after the inflation is over, the subsequent Friedmann expansion would take over with a curvature term which is totally negligible. Therefore the present state of the universe should be indistinguishable from the flat $k = 0$ model, and the present value of the density parameter Ω (the ratio of the true density to the critical density) should be very close to unity. A more reliable determination of Ω would provide a crucial test of the inflationary theory.

Alan Guth's original inflationary theory has some shortcomings and his theory has been modified by A. Linde, A. Albrecht, P. J. Steinhardt, and Guth himself. Even the new inflationary theory was not altogether successful. Theorists are still exploring the new model's implications and tinkering with its details, striving to fill in gaps and to patch up apparent flaws. But many physicists and astronomers think the inflationary theory is on the right track.

11-7 Cosmic strings: relics of the early universe?

We have seen that Alan Guth's inflation occurred when the universe went from the high-energy symmetric state of a false vacuum to the low-energy symmetry-broken state of a true vacuum. A question arises naturally: could a defect exist in the early breakdown of symmetry of the forces? Yes. Alexander Vilenkin first made such a proposal in 1977, and these defects are now called **cosmic strings**. This may be compared with the breakdown of symmetry when ice forms from water. Slight impurities cause the production of defects in the crystallized ice. Similarly, any slight randomness in the early, unified universe causes imperfections that show up as defects later on.

Cosmic strings are bits of isolated space containing very high energies. Perhaps cosmic strings might be better described as pure bits of unified force fields in which the strong and electroweak forces remain unified. They are incredibly narrow, only a few atoms wide, and may form long loops billions of kilometers long — thus the term strings. Although they are different from ordinary matter, the strings are cracks or distortions in space-time, so they produce an effect of great mass concentration. They have the same gravitational attraction as a cluster of galaxies wound up into an incredibly narrow filament and stretched out over billions of kilometers in space. J. Ostriker and Ed Witten have found an even more fascinating property of the strings. They may not just attract matter around them, they may also repel it. This is caused by their strong radiating tendency. This radiation can be intense enough to push away matter near the string, effectively clearing out a channel of space for billions of kilometers. If the cosmic strings do exist, they are an ideal candidate for the "missing mass" in the universe.

What evidence do we have of cosmic strings? In 1979 Robert Kirshner was studying the galaxies in the direction of the constellation Bootes. He discovered that over a certain range of redshifts (hence distances) no galaxies — single or in cluster — were seen at all. Such a region free of galaxies is called a **void**. Now many other cosmic voids had been found in other regions of space. Some were a fraction of the size of the Bootes void, while others were dozens of times larger. A curious aspect of these voids was that clusters of galaxies bordered the edge of the void in a way that showed a gigantic filamentary structure (Figure 9). The voids and filaments do not make sense with the standard big bang model. Some physicists suspect that cosmic strings may be responsible for these filaments.

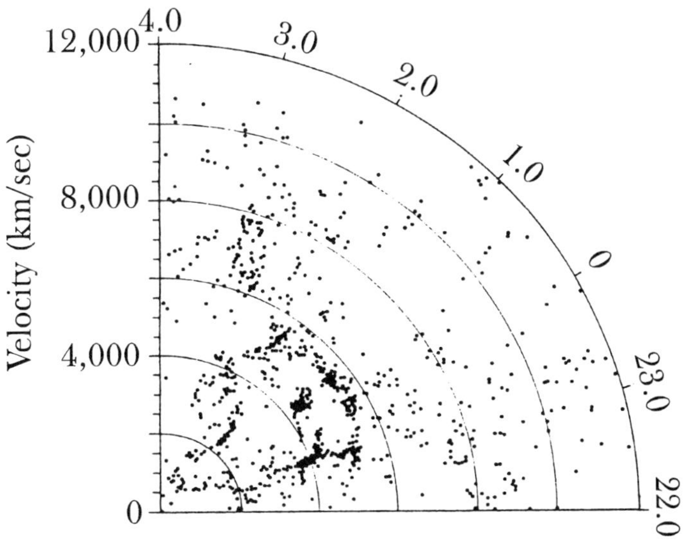

Figure 9: The large-scale distribution of galaxies.

In 1986 Alan Dressler and his colleagues discovered that our galaxy and its neighbors are streaming toward a distant point, roughly in the direction of the constellation Centaurus. Galaxies over nearly 100 million light years are cruising at speeds of 400 to 1000 km/s, and are converging toward a region roughly 150 million light years from our galaxy. They identified the cause of this galaxy streaming as the gravitational pull of a **Great Attractor**, an extensive continent of dark matter. Some physicists speculate that the Great Attractor may well be an invisible cosmic string.

Some physicists speculate that cosmic strings may also play an important role in forming galaxies. The early universe must have been exceedingly smooth, as revealed by the 3-K cosmic background radiation which is extremely isotropic. Galaxies could hardly have formed without sources of mass that served as seeds. What created these seeds in the first place? Some cosmic strings might be the seed for galaxy formation.

Although cosmic strings are very appealing as an explanation for voids, galaxy formation, Great Attractor, and so on, their influence poses complications in performing cosmological tests. For example, if the strings add streaming to galaxies, it may not be possible to separate the effects of streaming and galactic evolution from those caused by the universe's expansion.

Further Reading

Silk, J. *The Big Bang.* New York, W.H. Freeman and Company, 1989.

Weinberg, S. *The First Three Minutes.* New York, Basic Book, Inc., 1977.

Dirac, Paul A.M. *General Theory of Relativity.* New York, John Wiley & Sons, 1975.

Landau, L.D. & Lifshitz, E.M. *The Classical Theory of Fields* (Chapters 10 to 14), 4th edition; New York, Pergamon Press, 1975.

Sabbata, V.D. & Gasperini, M. *Introduction to Gravitation.* New York, World Scientific Publishing Co. Pte Ltd., 1985.

Weinberg, S. *Gravitation and Cosmology.* New York, John Wiley & Sons, 1972.

Narlikar, J.V., *Introduction to Cosmology.* New York, Cambridge University Press, 1983.

Index

About the Author

Tai L. Chow was born and raised in China. He received a BSc degree in Physics from the National Taiwan University (Taiwan, ROC), and a PhD from the University of Rochester (Rochester, NY, USA). Since 1976, Dr Chow has been in the Department of Physics at the California State University, Stanislaus. He has published more than 30 research papers, primarily in astrophysics. Dr Chow is also the author of the textbook *Classical and Modern Mechanics* (John Wiley, 1994).